9th International Conference on Environmental Science and Technology (ICEST 2018)

IOP Conference Series: Earth and Environmental Science
Volume 182

Prague, Czech Republic
20 – 22 June 2018

ISBN: 978-1-5108-7191-5
ISSN: 1755-1307

Printed from e-media with permission by:

Curran Associates, Inc.
57 Morehouse Lane
Red Hook, NY 12571

CURRAN ASSOCIATES INC. proceedings.com

Some format issues inherent in the e-media version may also appear in this print version.

Printed by Curran Associates, Inc. (2018)

For permission requests, please contact the Institute of Physics
at the address below.

Institute of Physics
Dirac House, Temple Back
Bristol BS1 6BE UK

Phone: 44 1 17 929 7481
Fax: 44 1 17 920 0979

techtracking@iop.org

Additional copies of this publication are available from:

Curran Associates, Inc.
57 Morehouse Lane
Red Hook, NY 12571 USA
Phone: 845-758-0400
Fax: 845-758-2633
Email: curran@proceedings.com
Web: www.proceedings.com

Table of contents

Volume 182

2018 9th International Conference on Environmental Science and Technology
20–22 June 2018, Prague, Czech Republic

Accepted papers received: 10 August 2018
Published online: 7 September 2018

Energy and Power Engineering

IOP Conf. Series: Earth and Environmental Science **182** (2018) 011001 doi:10.1088/1755-1315/182/1/011001

PREFACE

It is my great pleasure to welcome you to the 2018 9th International Conference on Environmental Science and Technology which was held in Faculty of Civil Engineering, Czech Technical University in Prague, Czech during 20-22 June, 2018. ICEST 2018 was dedicated to issues related to Environmental Science and Technology.

I, Prof. Roberto San Jose, from Technical University of Madrid (UPM), Madrid, Spain, held the keynote speech with title "A health impact assessment of traffic restrictions during Madrid NO_2 episode". Prof. Ing. Milan Holicky from Klokner Institute, Czech Technical University in Prague, Czech Republic had a keynote speech with title "Operational Procedures for Estimating Fractiles of Random Variables in Engineering and Science". Prof. Barry Jones from California Polytechnic State University, USA, held the keynote speech with title "The Failure to Deliver Infrastructure Projects on Time and to Budget – do we have the right tools and contract strategies". Prof. Bedřich Moldan from Charles University, Czech Republic performed the keynote speech with title "Impleentation of environmental science and technology: Opportunities and barriers".

There were five presentation sessions in which S.A. Osman, Tufan Tang, S.Maleki, Ryoichi Kanno and MR Usman have received the best presentations awards. Many researchers, engineers, academics as well as industrial professionals from all over the world have presented their research results and development activities to end the whole conference with a great success.

The major goal and feature of the conference was to bring academic scientists, engineers and industry researchers together to exchange and share their experiences and research results, and discuss the practical challenges encountered and the solutions adopted. Professors from Czech, Spain and USA were invited to deliver keynote speeches regarding latest information in their respective expertise. It has been a golden opportunity for students, researchers and engineers to interact with the experts and specialists to obtain their advice on technical matters as well as sales and marketing strategies.

The proceedings present a selection from papers submitted to the conference from universities, research institutes and industries. All papers were subjected to peer-review by conference committee members and international reviewers. The papers selected based on their quality and relevancy to the conference.

This volume presents to the readers the recent advances in the field of Environmental Science and Technology, Environmental dynamics, Meteorology, Hydrology, Geophysics, Atmospheric physics, Physical oceanography, Global environmental change, ecosystems management, Climate and climatic changes, Global warming, Ozone layer depletion, Carbon capture and storage, Biofuels, Integrated ecosystems

IOP Conf. Series: Earth and Environmental Science **182** (2018) 011001 doi:10.1088/1755-1315/182/1/011001

management, Satellite applications in the environment, Environmental restoration and ecological engineering, Habitat reconstruction, Biodiversity conservation, Deforestation, Wetlands, etc..

We would like to thank all authors who have contributed to this volume and also the organizing committee, reviewers, speakers, chairpersons, sponsors and all the conference participants for their support to ICEST 2018.

Prof. Roberto San Jose
Technical University of Madrid (UPM), Madrid, Spain
July 3, 2018

IOP Conf. Series: Earth and Environmental Science **182** (2018) 011002 doi:10.1088/1755-1315/182/1/011002

Peer review statement

All papers published in this volume of *IOP Conference Series: Earth and Environmental Science* have been peer reviewed through processes administered by the proceedings Editors. Reviews were conducted by expert referees to the professional and scientific standards expected of a proceedings journal published by IOP Publishing.

Screening of Solid Waste as Filler Material for Constructed Wetlands

Dina M R Mateus[1] and Henrique J O Pinho[2]

1 Technology, Restoration and Arts Enhancement Center (Techn&Art), Instituto Politécnico de Tomar, Portugal
2 Smart Cities Research Center (Ci2), Instituto Politécnico de Tomar, Portugal
E-mail: hpinho@ipt.pt

Abstract. The reuse of solid waste can contribute to reducing Earth's resource depletion, directly through use in the original production processes or by valorisation in alternative applications. In the present work, ten solid wastes were evaluated as candidates for filling material in constructed wetlands (CWs). For that purpose, physical characterization, leaching and adsorption tests were conducted. Limestone fragments and brick fragments resulting from construction activities, coal slags resulting from power plants, snail shells resulting from the food and catering industry, and cork granulates resulting from the cork industry have potential for use as CW fillers. These five materials have adequate physical properties and some capacity to adsorb phosphorous and organic compounds from wastewater. On the other hand, crushed eggshells resulting from egg farms, dealcoholized grape pomaces resulting from alcohol distilleries, olive seeds waste from olive-oil mills, and pine bark fragments and wood pellets resulting from forestry cleaning activities, wood mills and pulp mills did not demonstrate sufficient potential to be used as CW fillers, either because they have very low adsorption capacities or leach compounds in contact with water, or because they have less adequate physical properties. None of the tested solid wastes showed the ability to adsorb nitrogen compounds. Although the five selected materials do not present a special capability for adsorption of nitrogen, phosphorous and organic compounds, they can all be valued as CW fillers, representing a way to reduce the amount of solid waste sent to landfills.

1. Introduction

The objectives and strategies underlying the Sustainable Development paradigm are receiving increasing attention from the global community as a way to overcome the continuous increase in world population and the consequent pressure on available resources [1]. The concept of circular economy represents a vast set of possible actions that can contribute to sustainable development [2] [3]. One of the ways to implement the concept of circular economy is the reuse of waste in the same industrial or urban processes where they were originated, or in other processes in which they can be valued as raw materials or for energy conversion [4] [5]. In particular cases, waste can be used to treat waste, as for example in drinking water or wastewater treatment applications [6] [7].

Constructed wetlands (CWs) are an example of eco-efficient wastewater treatment technology and can be an application of the waste to treat waste concept [8]. Water purification processes by CWs are based on the assimilation of pollutants by macrophyte plants and the microbiological community, which contribute mainly to the assimilation of organic compounds [9]. In sub-surface flow CWs, the water under treatment flows through a porous filler material, for example sand. In this type of CW, the filler material serves as a support for plant growth and microbial community development. The filler materials, themselves, can contribute to the removal of water pollutants by physical and chemical

processes such as filtration, adsorption and precipitation [10]. The use of solid waste as filler materials can improve CW sustainability and represent a way to reduce landfill disposal of such materials.
The use of waste materials as CW fillers has already been studied by several researchers, namely for sorption of phosphorous [11]. This work focuses on solid waste generated locally to wastewater treatment facilities, to avoid transport costs, and to be used without any previous treatments, which may improve the sorption potential but require extra energy and material resources. Moreover, the present study is the first step to evaluate the future use of mixtures of these materials as CW fillers, combining their different sorption capacities for different pollutants and minimizing the possible drawbacks of some of them as fillers.

2. Materials and Methods

2.1. Materials

Ten solid waste materials were tested: clay brick fragments are a mineral waste originated from crashed or nonconforming construction clay bricks; coal boiler slags are a mineralized alumina-silicate waste resulting from coal power plants; cork granulates are a lignocellulosic waste resulting from cork processing industries; crushed eggshells are a mineral waste resulting from egg farms; dealcoholized grape pomaces are a lignocellulosic waste material resulting from wine alcohol distilleries; Moleano's rock limestone fragments are a mineral waste originated during the cutting and laying of stone tiles, slabs and similar construction elements used in buildings; olive seeds are a lignocellulosic waste from olive-oil production; pine bark fragments are a lignocellulosic waste resulting from forestry cleaning activities, wood mills and pulp mills; snail shells are a mineral waste from catering and food processing industries; wood pellets are a mechanical aggregate of waste pine wood resulting from forestry cleaning activities and wood mills.

2.2. Methods

The particle size distribution of the studied waste solids was examined using standard sieve analysis techniques and the values of d_{10} and d_{60} were determined [12]. The pycnometer method was used to evaluate the true density of the solids, according to European Standards [13]. The loose bulk density and voids were determined, also following European Standards [14].
To evaluate the possibility of nutrient leaching from the waste materials, a set of tests was carried out. All tested materials were first washed with tap water and dried at 60 °C until constant weight. About 10 g of each material was contacted with 150 ml of tap water in an Erlenmeyer flask for 40 h on an orbital shaker (agitation speed of 200 rpm). After that period, the liquid phase was separated from the solids by filtration and analysed to determine the total phosphorus (TP), total nitrogen (TN) and chemical oxygen demand (COD). pH and conductivity were also assessed.
To evaluate the adsorption potential, samples of the tested waste materials were contacted with an aqueous solution of KH_2PO_4 (ISO, pro analysis) and KNO_3 (ISO, pro analysis), to represent sources of inorganic phosphorous and nitrogen, respectively. To simulate the adsorption of organic compounds, toluene (ISO, pro analysis) was added to the solution. Toluene was chosen as a representative of organic water pollution as it is included on the BTEX group of pollutants. The solution was composed of 19.49 ± 0.81 mg/L of P, 18.68 ± 0.73 mg/L of N and 290 ± 3 mg/L of toluene. About 10 g of each solid material was contacted with 150 mL of the P-N-toluene solution in Erlenmeyer flasks for 24 h on an orbital shaker (agitation speed of 200 rpm). After that period, the liquid phase was separated from the solids by filtration and analysed to determine the total phosphorus, total nitrogen and COD content, pH and conductivity.
The materials that showed the best properties for use as constructed wetland filling were used to evaluate adsorption isotherms. To accomplish that goal, samples of the materials were contacted separately with aqueous solutions of potassium dihydrogen phosphate and aqueous solutions of toluene. About 5 g of solid samples were contacted with 75 mL of the solutions, one set of phosphate solutions of total phosphorus concentration in the range of 2 to 100 mg_P/L, and other set of toluene solutions with COD in the range of 50 to 800 mg_{O2}/L (13.0 – 261 mg/L toluene). The Erlenmeyer flasks containing the solid samples and the aqueous solutions were incubated during 24 h in a

temperature controlled orbital shaker (agitation speed of 200 rpm and temperature of 22 ºC). After that period, the liquid phase was separated from the solids by filtration and analysed to determine the TP and COD content. The adsorbed mass of solute was evaluated from a mass balance to the aqueous phase. Langmuir adsorption isotherm model (eq. 1) was fitted to the experimental data to estimate the maximum adsorption capacity of the materials (W_{max}) for phosphorus as TP and for organic compounds as COD. The fittings were performed by non-linear regression using the IBM's SPSS software, version 25.

$$W_e = W_{max} \frac{KC_e}{1+KC_e} \quad (1)$$

W_e in equation 1 represents the solute adsorbed on the solid (mg/g) and C_e represents the equilibrium concentration of the solute in the aqueous solution (mg/L). W_{max} (mg/g) and K (L/mg) are the Langmuir model parameters associated to the maximum adsorption capacity and the solute affinity for the adsorbent, respectively.
TP, TN and COD analysis were performed with reagent kits from Hanna Instruments. A COD heat block (HI-839800, Hanna Instruments) was used to perform the required digestions and a photometer (HI-83399, Hanna Instruments) was used to perform the analysis. At least two replicates were performed for all assays and measurements.

3. Results

3.1. Physical Properties of the Tested Materials

Table 1 presents the physical properties obtained experimentally for the tested solid waste materials. It was not possible to use the sieve technique for the crushed eggshells and dealcoholized grape pomaces materials. Grape pomace is a very heterogeneous material composed of particles of irregular shapes. The eggshell contains remains of the egg's liquid part, which makes the material sticky. For these reasons, both materials do not freely cross the sieves.

Table 1. Physical properties of the tested solid waste.

Solid waste	Density (Mg/m^3)	Bulk density (Mg/m^3)	d_{10} (mm)	d_{60} (mm)	d_{60}/d_{10}
Clay brick fragments	2.67 ± 0.03	1.069 ± 0.003	1.9	6.5	3.4
Coal slags	2.10 ± 0.05	0.880 ± 0.004	0.06	1.2	19
Cork granulates	0.15 ± 0.01	0.072 ± 0.001	2.2	3.6	1.6
Crushed eggshells	2.53 ± 0.02	NA	NA	NA	NA
Grape pomaces	1.21 ± 0.01	0.280 ± 0.003	NA	NA	NA
Limestone fragments	2.69 ± 0.01	1.309 ± 0.001	7.2	11.2	1.6
Olive seeds	0.88 ± 0.01	0.586 ± 0.001	1.2	2.8	2.3
Pine bark	0.38 ± 0.01	0.137 ± 0.003	4.8	24.2	5.1
Snail shells	2.56 ± 0.06	0.130 ± 0,001	11.2	14.8	1.3
Wood pellets	1.06 ± 0.01	0.194 ± 0.001	0.28	0.90	3.2

NA = Not available

The d_{10} and d_{60} parameters are the diameters corresponding to 10% and 60% finer in the particle size distribution, by weight, respectively. The ratio d_{60}/d_{10} is the uniformity coefficient, which should be less than 4 to prevent the risk of water flow clogging [15]. Materials with a coefficient of uniformity of more than 4 are not recommended for use in CW unless they are mixed with other materials. This is the case for the coal slags and pine bark.

3.2. Nutrient Release Experiments

Table 2 presents the results of the evaluation of nutrient release by the solid waste. It was found that, in general, the studied materials do not release significant quantities of phosphorus. Among them, inorganic materials showed the least ability to release phosphorus compounds. Snail shells and eggshell residues presented slightly higher values than other inorganic materials, a result which may be explained by the fact that they may contain remains of organic materials. The cork granulates, pine bark, wood pellets, olive seeds and grape pomaces liberate more significant quantities of phosphorus compounds, which may be explained by their organic nature.
All the materials studied release only small quantities of nitrogen compounds. Limestone and brick fragments release very small amounts. The eggshell, wood pellets and grape pomaces liberate higher amounts, but not more than 0.25 mg/g.
Concerning the potential for releasing compounds that contribute to COD, the brick fragments, limestone fragments, snail shell and cork granulate have the lowest values, less than 1 mg/g. Coal slags have a slightly higher value of 1.3 mg/g. The remaining materials present values higher than 5 mg/g. The wood pellets, olive seeds and grape pomaces showed high potential to release compounds that contribute to COD and therefore are not suitable as filling material to be used in CW for wastewater treatment.

Table 2. Potential of nutrients release from the materials upon contact with tap water.

Solid waste	TP (mg_P/g)	TN (mg_N/g)	COD (mg_{O2}/g)	pH	Conductivity (μS/cm)
Tap water	0	0	0	7.96 ± 0.30	152 ± 25
Clay brick fragments	ND	0.004 ± 0.004	0.75 ± 0.42	7.37 ± 0.42	153 ± 21
Coal slags	ND	0.021 ± 0.029	1.61 ± 0.42	7.22 ± 0.43	151 ± 50
Cork granulates	0.016 ± 0.002	0.026 ± 0.003	1.62 ± 0.74	6.95 ± 0.12	203 ± 52
Crushed eggshells	0.031 ± 0.002	0.128 ± 0.052	3.65 ± 0.44	7.55 ± 0.25	461 ± 110
Grape pomaces	0.028 ± 0.039	0.250 ± 0.054	46.1 ± 1.6	3.9 ± 0.1	3420 ± 42
Limestone fragments	ND	0.006 ± 0.001	1.00 ± 0.50	7.55 ± 0.43	113 ± 64
Olive seeds	0.027 ± 0.04	0.026 ± 0.012	11.3 ± 2.7	6.31 ± 0.15	301 ± 35
Pine bark	0.008 ± 0.003	0.076 ± 0.033	6.6 ± 2.3	4.46 ± 0.53	199 ± 12
Snail shells	0.004 ± 0.006	0.042 ± 0.034	1.27 ± 0.87	7.35 ± 0.58	187 ± 18
Wood pellets	0.008 ± 0.011	0.066 ± 0.071	25.8 ± 2.2	4.79 ± 0.04	297 ± 59

ND = Not detected

Table 2 also presents the pH and conductivity of the water, before and after contact with the tested materials. These two parameters should also be considered for the selection of materials to be used as CW fillers. It has been found that all materials tend to decrease the water pH. This reduction is not significant for most materials, particularly for inorganic materials and for the cork granulate. For the other materials, especially for grape pomaces, pine bark and wood pellets, a more significant reduction of pH was observed. The pH reduction is not usually advantageous for the proper functioning of the CW, but may be useful in the case where the CW are used to treat alkaline wastewater.
The increase in the water electrical conductivity by the contact with the materials allows to infer about the possibility of ion release to the aqueous phase [16]. The release of calcium, aluminium, iron and magnesium ions may favour the removal of phosphorus through the formation of insoluble precipitates, but the release of other ions, especially those from heavy metals, are disadvantageous for water treatment. The tested mineral materials and granulated cork did not present a tendency to release ions. Eggshell and the tested organic materials lead to an increase in the water conductivity, particularly the grape pomaces.
Based on the results obtained, it was concluded that eggshell residues are not suitable for use as CW filler because of the high organic load released. However, the disadvantage pointed out can be overcome if the organic residues present in the eggshells are previously removed by calcination, for

example. However, the eggshells pre-treatment can be costly, which reduces the attractiveness of this solution. Grape pomace released phosphorus, nitrogen and COD compounds and led to a significantly acidic pH, so it is also not a good choice. Wood pellets have proved to be a source of nitrogen and COD and cause a significant reduction of pH. This material tends to release wood fibres into the water, which can lead to clogging problems. In view of these observations, the three mentioned materials were not used in the adsorption studies.

3.3. Preliminary Adsorption Experiments

The results of the preliminary adsorption tests are presented in Table 3. It can be seen from the results that none of the materials showed capacity for nitrogen adsorption. Moreover, no adsorption of TP, TN or COD was detected for the olive seeds and pine bark.

Table 3. Preliminary essays to evaluate the potential of adsorption from aqueous solutions.

Solid waste	TP (mg_P/g)	TN (mg_N/g)	COD (mg_{O2}/g)
Clay brick fragments	0.035 ± 0.049	ND	1.79 ± 0.47
Coal slags	0.027 ± 0.002	ND	1.58 ± 0.13
Cork granulates	ND	ND	1.30 ± 0.27
Limestone fragments	0.034 ± 0.016	ND	1.83 ± 0.11
Olive seeds	ND	ND	ND
Pine bark	ND	ND	ND
Snail shells	0.018 ± 0.004	ND	0.58 ± 0.15

ND = Not detected

Limestone fragments, clay brick fragments and coal slags revealed some phosphorus adsorption capacity. The snail shells showed a lower adsorption than the other mineral materials. The remaining materials did not show adsorption capacity for phosphorus compounds. It was observed that most of the materials showed a tendency to adsorb toluene, through the reduction observed in COD content. The potential of the materials to remove COD content from water followed the order limestone>brick>slags>cork>snail shells.

From this set of tests, it was concluded that the limestone fragments, clay brick fragments, coal slags, cork granulates and snail shells may have potential application as CW filler material. On the other hand, olive seeds and pine bark showed a lower potential, so they were not used for the more complete adsorption studies.

3.4. Adsorption Isotherm Experiments

Following the nutrients release and preliminary adsorption tests results, five materials were selected for more extended adsorption essays. The adsorption capacity for phosphorus compounds and for oxidable compounds was studied using different concentrations of phosphate and toluene solutions. The adsorption isotherm of nitrogen compounds was not evaluated, since none of the materials has showed potential to remove TN in the preliminary adsorption experiments.

Figures 1 and 2 show the isotherms obtained by non-linear adjustment for phosphate and toluene adsorption, analysed as TP and COD, respectively. The figures show a high data scattering, which may be due to the irregular particle shape and dimensions, and to the heterogeneous composition of the waste, despite the efforts to use representative and similar samples.

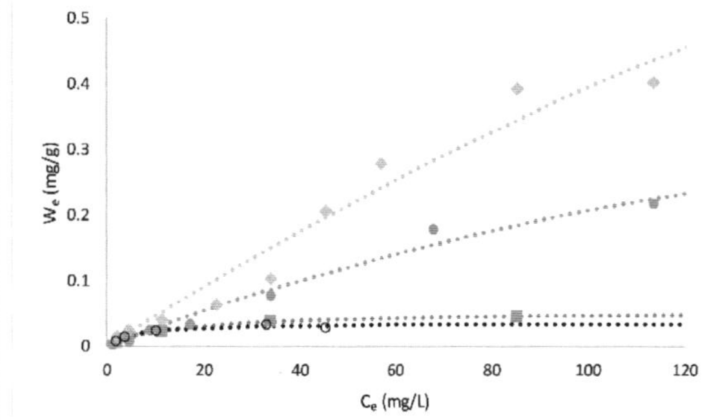

Figure 1. Fitting of Langmuir isotherm to TP adsorption data: ● clay brick fragments; ■ coal slags; limestone fragments; O snail shells.

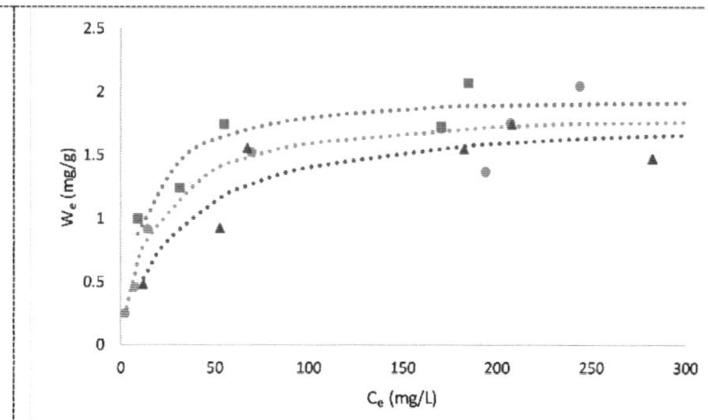

Figure 2. Fitting of Langmuir isotherm to COD adsorption data: ● clay brick fragments; ■ coal slags; ▲ cork granulates.

Table 4 contains the parameters obtained for the adjustment of the Langmuir isotherm to the experimental data. The potentiality for TP and COD removal can be related to the parameter W_{max}.

Table 4. Langmuir isotherm parameters for phosphate and toluene adsorption, analysed as TP and COD, respectively.

Solid waste	Phosphate adsorption (as TP)			Toluene adsorption (as COD)		
	W_{max} (mg/g)	K (L/mg)	r^2	W_{max} (mg/g)	K (L/mg)	r^2
Clay brick fragments	0.70 ± 0.47	0.004 ± 0.005	0.984	1.89 ± 0.41	0.054 ± 0.033	0.918
Coal slags	0.052 ± 0.009	0.069 ± 0.026	0.988	2.02 ± 0.75	0.081 ± 0.083	0.838
Cork granulates	ND	ND	ND	1.83 ± 0.36	0.032 ± 0.047	0.831
Limestone fragments	2.31 ± 0.65	0.002 ± 0.003	0.955	ND	ND	ND
Snail shells	0.035 ± 0.007	0.198 ± 0.094	0.961	ND	ND	ND

ND = Not determined; r^2 = coefficient of determination of the non-linear regression.

For the limestone fragments and snail shells the COD data showed a significant uncertainty, resulting in adsorption values close to zero. These results did not correspond to the expected, given the results obtained in the preliminary studies. However, these materials are predominantly composed of calcium carbonate, so it would not be expected that they adsorb organic materials well. Brick fragments, coal slags and cork granulates demonstrated organic matter adsorption capacities, evaluated in COD terms, very similar to each other and of the order of 2 mg_{O2} per g of solid material.
Regarding the adsorption of phosphorus compounds, the results obtained for the cork granulates confirmed the reduced potential of this material. On the other hand, the limestone fragments showed the greatest capacity for the adsorption of TP. The brick fragments had a capacity four times lower compared to limestone fragments, but still much higher than the organic waste materials. The results agree with the mechanisms usually indicated to justify the removal of phosphorus compounds in the CWs, which mainly consist of physical and chemical adsorption of phosphates.

4. Conclusions

Some waste materials can be used as waste-to-treat-waste and can be used as CW fillers, contributing to the treatment of wastewater through adsorption and associated sorption processes.
Among the various residual materials studied in this work, which are mainly deposited in landfills, limestone fragments and brick fragments resulting from construction activities, coal slags resulting from power plants, snail shells resulting from the food and catering industry, and cork granulates resulting from the cork industry have potential to be used as CWs fillers. These materials have

adequate physical properties and some capacity to adsorb phosphorous and organic compounds from wastewaters. The limestone fragments presented a relevant phosphorus adsorption capacity. The brick fragments had some potential to adsorb compounds of phosphorus and organic compounds. Coal slags showed the potential to adsorb organic compounds and, to a lesser extent, phosphorus compounds. The cork granulates showed the ability to adsorb organic compounds. The snail shells are, among these materials, the one with the least potential, as they showed a very low capacity to adsorb phosphorus compounds.

The remaining materials studied did not demonstrate potential to be used as CW fillers, because they have very low adsorption capacities, leach nutrients or other pollutants in contact with water, or because they have less adequate hydraulic properties.

None of the materials tested showed the ability to adsorb nitrogen compounds, but this nutrient is not usually removed by adsorption processes, so this does not limit the use of the materials as CW fillers.

Future work should involve the mixture of materials, taking advantage of their combined properties. Studies on the capacity of the materials for the establishment of microbial communities and for the development of macrophyte species, to promote biological removal processes, should also be carried out.

Although the studied materials did not present exceptional capacities of adsorption of nitrogen, phosphorous and organic compounds, all can be valued as CW fillers, which represents a way to reduce the amount of solid waste sent to landfills. The use of waste materials as CW fillers is an application of the concept of circular economy, replacing engineered filling materials that may present better removal efficiencies but have high production cost, waste generation and high energy consumption processes. The results obtained justify further tests in pilot-scale CWs.

5. References

[1] United Nations General Assembly on 25 September 2015 *Transforming our world: the 2030 Agenda for Sustainable Development*. Retrieved from http://www.un.org/ga/search/view_doc.asp?symbol=A/RES/70/1&Lang=E Acessed on: 17 jul 2017

[2] Ghisellini, P, Cialani C and Ulgiati S 2016 A review on circular economy: The expected transition to a balanced interplay of environmental and economic systems *J. Clean. Prod.* 114 11-32

[3] Winans K, Kendall A and Deng H 2017 The history and current applications of the circular economy concept *Renew Sustain Energy Rev* 68 825-33

[4] Maina S, Kachrimanidou V and Koutinas A 2017 A roadmap towards a circular and sustainable bioeconomy through waste valorization *Curr. Opin. Green Chem* 8 18-23

[5] Malinauskaite J, Jouhara H, Czajczynska, D *et al.* 2017 Municipal solid waste management and waste-to-energy in the context of a circular economy and energy recycling in Europe *Energy* 141 2013-44

[6] Grace M A, Clifford E and Healy M G 2016 The potential for the use of waste products from a variety of sectors in water treatment processes *J. Clean. Prod.* 137 788-802

[7] Saxena A, Bhardwaj M, Allen T, Kumar S and Sahney R 2017 Adsorption of heavy metals from wastewater using agricultural–industrial wastes as biosorbents *Water Sci.* 31 189-97

[8] Mateus D M R, Vaz M M and Pinho H J O 2012 Fragmented limestone wastes as a constructed wetland substrate for phosphorus removal *Ecol. Eng.* 41 65-9

[9] Wang M, Zhang D Q, Dong J W and Tan S K 2017 Constructed wetlands for wastewater treatment in cold climate — A review *J. Env. Sci.* 57 293-311

[10] Wu J, Xu D, He F, He J and Wu Z 2015 Comprehensive evaluation of substrates in vertical-flow constructed wetlands for domestic wastewater treatment *Water Practice Tech.* 10 625-632

[11] Wu H, Zhang J, Ngo H H, Guo W, Hu Z, Liang S, Fan J and Liu H 2015 A review on the sustainability of constructed wetlands for wastewater treatment: Design and operation *Bioresource Technol.* 175 594-601

[12] European Committee for Standardization EN 933-1:2012 *Tests for geometrical properties of aggregates - Part 1: Determination of particle size distribution - Sieving method*

[13] European Committee for Standardization EN 1097-6:2013 *Tests for mechanical and physical properties of aggregates - Part 6: Determination of particle density and water absorption*
[14] European Committee for Standardization EN 1097-3:1998 *Tests for mechanical and physical properties of aggregates - Part 3: Determination of loose bulk density and voids*
[15] Arias C A, Del Bubba M and Brix H 2001 Phosphorus removal by sands for use as media in subsurface flow constructed reed beds *Water Res.* 35 1159-68
[16] Wang Z, Dong J, Liu L, Zhu G and Liu C 2013 Screening of phosphate-removing substrates for use in constructed wetlands treating swine wastewater *Ecol. Eng.* 54 57 – 65.

Acknowledgments

This work was supported by Program FEDER, ref. POCI-01-0145-FEDER-023314, project VALORBIO. The authors acknowledge the collaboration of the Lab.IPT staff and the assistance of students of chemical and environmental technology courses held at Instituto Politécnico de Tomar. Special thanks to Alcino Serras, Ana Alves, Isabel Silva, Joel Nunes and Nuno Graça.

Study on the Short-term Treatment Technology of Landfill Leachate

Jin Li[1],FengnanLiu[2], Feng Ding[3], Xueren Dong[1*], Qinqin Xu[4]

1 School of Mechanical Engineering, University of Jinan, Jinan, Shandong, 250022, China
2 Safety Productionand Supervision Bureau of Yishui, Linyi, Shandong, 276400, China
3 Inspection and Testing Center of Yishui,Linyi, Shandong, 276400, China
4 School of Science, Nanjing University of Technology, Nanjing, Jiangsu, 210094, China
E-mail: ljawyjn@163.com

Abstract. Domestic garbage is mainly treated by sanitary landfill.The main characteristics of landfill leachate produced in the sanitary landfill process are high concentration of COD and BOD, high ammonia nitrogen content,high levels of refractory organics,heavy metal ions, poor biodegradability, unstable water quality, difficult handling and so on. If the treatment is not carried out, the environment will be severely damaged. According to the needs of the project, this paper proposes a set of short-term landfill leachate treatment process, which involves ozone oxidation, internal electrolytic oxidation technology, coagulation precipitation, deep oxidation of the chain reaction, microwave field catalytic oxidation, adsorption and other treatment processes. Eliminating the biochemical process, the entire process flow is relatively short, and the processing efficiency is high. And the use of PLC-based control system to achieve automatic control of landfill leachate treatment. Since the leachate water quality has a wide range of variation, the PID control method based on BP neural network adaptive control strategy is adopted to achieve on-line adjustment of PID control parameters. The final effluent water quality can meet the emission standards.

1. Introduction

With the increase of the urban population, the expansion of the city scale and the improvement of the residents' living standards in China, the production of urban domestic waste in China has increased dramatically. At present, waste disposal methods include sanitary landfills, waste incineration, and composting. The main reason for adopting the sanitary landfill method in China is that the method is not only low in cost, but also able to handle a large amount of garbage, and it is also relatively simple to manage and has strong adaptability [1].

In the urban landfill process, factors such as atmospheric precipitation, surface runoff, and self-contained water in the garbage will cause the pollutants in the garbage to dissolve out together with the water, thereby forming landfill leachate. Leachate has more than 20 toxic and hazardous pollutants that are preferentially controlled by the environment. Not only is the composition of the leachate complex, but also the water volume and water quality are very volatile. These characteristics make leachate treatment a worldwide issue [2]. Leachate will seriously affect people's health and living environment, and the harmless treatment of leachate is more and more important. Therefore, it is of great significance to study the leachate treatment process [3].

IOP Conf. Series: Earth and Environmental Science **182** (2018) 012002 doi:10.1088/1755-1315/182/1/012002

Domestic and foreign landfill leachate treatment methods include physicochemical methods, biological methods, land treatment methods, and some combination processes. Physicochemical treatment has a good effect on the leachate with poor biological treatment effect. The change of water quality and water quality has little effect on it, and the effluent water quality is relatively stable.Physical and chemical reactions generally include coagulation and sedimentation, activated carbon adsorption, chemical oxidation, ion exchange, membrane separation, and catalytic oxidation. Biological methods can be divided into aerobic, anaerobic, and anaerobic-aerobic combi-nation processes. The principle of the land treatment method is to use the natural self-cleaning ability of the earth to treat the organic matters in the leachate through processes such as adsorption, filtration, oxidation, or microbial utilization. Currently, there are few applications. Some studies have used microwave and activated carbon in combination to enhance persulfate treatment of landfill leachate. The results show that the removal rate of COD and ammonia in landfill leachate is 78.2% and 67.2%, and the ratio of BOD_5/COD is increased from 0.17 to 0.38[4]. The experimental results show that the removal rates of COD, ammonia nitrogen and chromaticity are 88%, 79% and 100%, and BOD_5/CODfrom 0.07 to 0.34 [5]. Some studies used aluminumsulfate and ferric chloride to treat a stable leachate, and the COD removal rate was over 75% [6].In some studies, coagulation combined with photo-Fenton's method has been used to treat aged landfill leachate. It was found that the introduction of coagulation in the photo-Fenton combined treatment of landfill leachate compared to the single use of the photo-Fenton process for leachate treatment. The process can increase the COD removal rate from 63% to 89%, and the treated leachate exhibits no toxicity [7]. This dissertation mainly discusses a short-range landfill leachate treatment process. The leachate is mainly treated by physical treatment.

2. Landfill Leachate Treatment Process

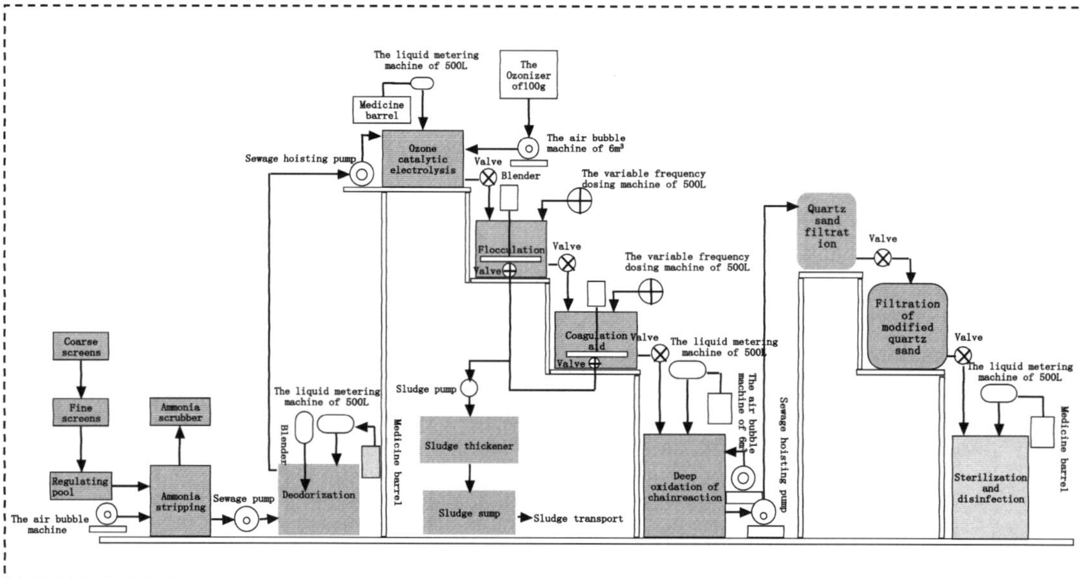

Figure1. Treatment process of landfill leachate

The treatment process for the formulated landfill leachate is shown in figure 1. The following describes each process in detail.

2.1. Coarse grid

The coarse grille is placed before the intake pump room. It is used to remove large-size floating andsuspended materials and protect the operation of the pump to avoid accidents such as impeller winding and plugging, and ensure that the subsequent processing facilities can operate normally. The rotary grid decontamination machine is generally composed of rake teeth installed at a certain distance on the slewing chain. Driven by the driving device, the revolving chain drives the rake tooth to rotate in a certain direction.In the water surface, the rake teeth are moved from the bottom to the top to fish out the floaters in the water, rotate to the top and then remove the floaters. The filtering accuracy is 20mm [8].

2.2. Fine grid

Fine grids are set behind the coarse grids, because there are various fine linear floats, flake-like floats, such as hair, fabric fines, plastics and rubber fragments in the landfill leachate. With the accumulation of these fibrous substances, the sludge will be entrapped in the fiber mass and the sludge will be lost with the effluent.The drum-type fine grid is used to filter the landfill leachate. The filtration accuracy is 5mm [9].

2.3. Regulating pool

The garbage composition of the landfill site is complex, and the pH of the leachate generally fluctuates within the range of 4-9. The pH value will have a greater impact on subsequent treatment processes. In the ammonia stripping process, the most suitable pH value is 10.5 [10]. Alkaline lime water is added to the conditioning tank through the dosing machine, and the pH of the leachate is adjusted to 10.5.

2.4. Ammonia stripping

The basic principle of the blow-off method is to use air to blow off under the alkaline condition by utilizing the difference between the actual concentration of the volatile components such as ammonia nitrogen contained in the waste water and the equilibrium concentration under the determined conditions.In order to improve the stripping efficiency, recover the useful gas and prevent the secondary pollution, the packed tower is adopted.The packing tower is constructed by gas-liquid contacting device, and a certain height of filling layer is set in the tower[11]. The lime-treated leachate is showered from the upper part of the tower to the filling to form water droplets. It falls down in the gap between the fillers, and then the air is blown upward from the bottom of the tower with a fan to make ammonia evolve into gas to escape from the water. The ammonia-containing tail gas enters ammonia absorption tower. After the leachate is blown off by ammonia, not only a large amount of free ammonia is removed, but also some phenols, cyanides, sulfides, and other difficult-to-biodegrade volatile substances are removed [10].

2.5. Deodorant

The filtered leachate was passed through the deodorant tank, and sodium hypochlorite was added while the mixer was stirring. Simultaneously, sulfuric acid was added to adjust the leachate to acidity.In terms of the properties of various oxidants, sodium hypochlorite has good effect and economic efficiency, so it is widely used.In the solution, sodium hypochlorite is present in the hypochlorous acid situation. The source of odor is mainly hydrogen sulfide, organic sulfur, volatile fatty acids, etc. Sodium hypochlorite has strong oxidizing properties and can kill bacteria and viruses. At the same time, it can remove hydrogen sulfide and so on. so it can deodorization [12].

2.6. Ozone catalytic electrolysis

Ozone oxidation technology is often used to remove refractory organics and chromaticity in leachate. Ozone oxidation alone used to treat landfill leachate has problems such as low ozone utilization and long reaction time. Therefore, a new combination of catalytic technology and ozone oxidation is used.On the basis of this, internal electrolytic oxidation method is used to produce oxygen radicals and other active groups to oxidize and degrade the organic matter in the leachate, and at the same time, the flocculation precipitation removes most of the heavy metals[13].First add iron-carbon filler to the tank, then inject ozone into the tank through the bubble machine, and add potassium monopersulfate inside by using the dosing machine. The sulfate radical ($SO_4^{-}\bullet$) produced by the activated persulfatecan rapidly organic pollutants and mineralize them into $CO_2$and inorganic acids.Compared with the traditional advanced oxidation technology, sulfate radicals are highly selectiveand low requirements for the external environment [14].The leachate is electrolyzed using a 1.2V potential difference generated by the iron carbon material filled in the leachate. Numerous micro-battery systems are formed in the reaction tank, and an electric field is formed in the action space.The new ecology [H], Fe^{2+}, hydroxyl radicals, sulfate radicals and other energy generated during the treatment process can undergo redox reactions with many components of the leachate. These oxidation reactions can destroy the chromophoric group or the chromophoric group of the colored substances in the leachate, or even break the chain, to achieve the effect of decolorization.The generated Fe^{2+} is further oxidized to Fe^{3+}, and their hydrates have strong adsorption-flocculation activity.The process can effectively remove the organic matter, heavy metal and chromaticity in the leachate, which can reduce the COD value of water quality and its toxicity.

2.7. Flocculation aid

The flocculant PAC was added by means of a variable liquid metering dosing machine. The aqueous solution of PAC is a hydrolytic product between aluminum chloride and aluminum hydroxide, and has a colloidal charge. Therefore, it has strong adsorption for suspended solids in water[13]. PAC is a kind of polymer inorganic flocculant.Polyaluminumhydrolyzate has polynuclear hydroxyl group containing Al^{3+}complex.These complexes are further hydrolyzed and the hydrolyzed product have the ability to neutralize the black colloid negative charge, compress the electric double layer, and reducethezeta potential of the colloid [15].

After the PAC flocculant was put into the leachate, the colloids were reduced or eliminated due to the potential, and the stability of the particles was destroyed. The destabilized particles aggregate with each other into larger particles. Flocculation sedimentation not only removes fine suspended particles in leachate, but also removecolor, oil, microorganisms, nitrogen, heavy metals and so on. The precipitated sludge enters the sludge thickening tank through the sludge pump.

2.8. Coagulation aid

After 7 minutes of addition of PAC flocculant, polyacrylamide (PAM) was added to the reaction tank via a frequency meter dosing machine. PAM is a macromolecule substance added to regulate or improve flocculation conditions and promote aggregation. It has been widely used in the field of water treatment[14]. The PAM molecular chain contains a certain number of polar groups, which can adsorb solid particles suspended in the waste water to bridge the particles or neutralize the particles to form large flocs. The density and weight of PAM can accelerate the sedimentation of particles in the suspension[16].The resulting sludge enters the sludge thickening tank through the sludge pump.

2.9. Chain reaction deep oxidation

Sodium persulfate is added to the barrel through the liquid metering dosing machine, and then bubbles of ozone are bubbled into the tank through the bubble machine. The chain-catalyzed multi-oxidation radical mutual catalytic deep-oxidation oxidant is composed of an active radical source, an active activator, a balance stabilizer, and a sustained-action accelerator, etc. It is a strong oxidation technology formed by the integrated excitation of electromagnetic field, microwave and ultraviolet radiation to form S- free radical as the core, hydroxyl radical (OH•), halogen free radical (Cl•), hydrogen peroxide free radical (HOO•) and so on. Microwave activation is a heating method that is different from thermal energy at the molecular level.It selectively heats absorbing materials through microwaves, catalyzes at low temperature, and penetrates quickly. The organic matter in landfill leachate undergoes violent catalytic and physicochemical reactions under the combined action of activated carbon and microwaves, and which is converted into insoluble substances or gases separated from water. Some of the organic pollutants are broken down under the action of microwave catalysis. They are decomposed into small molecules and are combined with activated carbon to form floc. Metal ions directly combine with activated carbon to produce sink flocculant precipitates. Ammonia nitrogen is converted into ammonia gas. When the concentration exceeds the standard, it can be absorbed and removed by a subsequent absorption device. Phosphorus in the leachate is converted to insoluble phosphate precipitates for removal. The treatment process has effective removal rates for

IOP Conf. Series: Earth and Environmental Science **182** (2018) 012002 doi:10.1088/1755-1315/182/1/012002

COD, NH_3-N, animal and vegetable oils, phosphates, LAS,Cr^{6+}, CN^-, Cu^{2+}, Zn^{2+}, Pb, Ni, total chromium and other pollution factors, especially for COD.

2.10. Quartz sand filtration
The purified quartz sand filter material is filled into the reaction tank. When the influent water flows through the filter layer from top to bottom, the suspended matter and the viscose particles in the leachate are removed, so that the turbidity of the leachate is reduced. This process can effectively remove suspended solids, organic compounds, colloidal particles, microorganisms, chlorine, smell and some heavy metal ions and other in leachate [17].

2.11. Modified quartz sand filtration
The reaction tank is filled with refined modified quartz sand filter material, and the graphene is combined with the quartz sand chemically to form a strong coating on the surface of the quartz sand. At the same time, quartz sand and graphene are combined to avoid the uncontrollability of nanomaterials in subsequent processing, and it is easier to recycle and reuse. Graphene forms a dense fiber filter layer that maximizes the use of nanoscalegraphene properties and it greatly improves the strength and toughness of composites. Graphene has a large theoretical specific surface area, and graphene evenly penetrates into the gap of the screen substrate, so that the filter fiber component has a large specific surface area and has a lot of microporous structures on it. Therefore, modified quartz sand has stronger adsorption effect than traditional filter materials. The process further filters out leachate suspended solids, organic matter, colloidal particles, microorganisms, chlorine, smell and some heavy metal ions. It has the characteristics of short adsorption time, high adsorption efficiency and long filter screen usage time[17].

2.12. Sterilization and disinfection
Potassium bisulfate is added to the tank by a liquid dosing machine. Potassium bisulfate compound salts are dissolved in water to release reactive oxygen speciesand high-energy, high-activity small-molecule radicals, neo-atom oxygen, chlorine radicals, hydroxyl radicals (OH•), sulfuric acid radicals (SO_4^-•) and a variety of active ingredientsare generated by high energy activator via chain reaction. So it becomes an efficient oxidation disinfectant, killing bacteria, spores, viruses, fungi and other microorganisms. At the same time, lime water was added to the tank to adjust the effluent pH to 7.

3. Control System

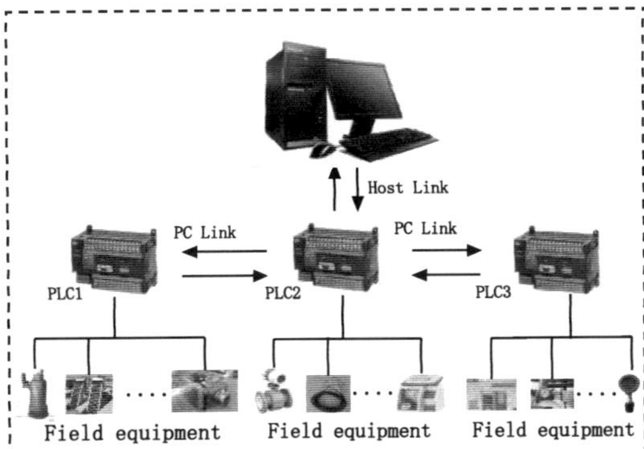

Figure2. Structure diagram of control system

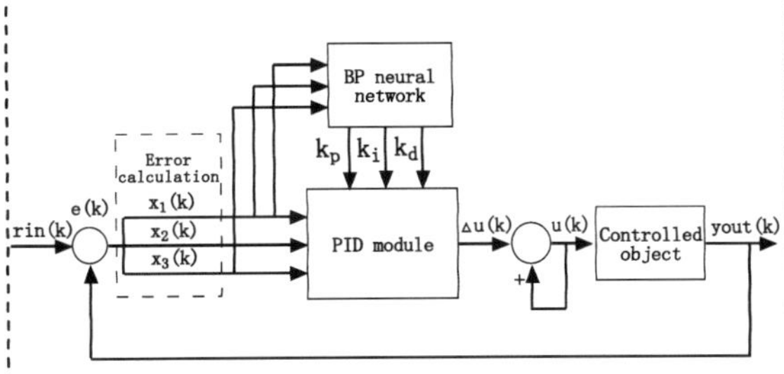

Figure3.BP-PID structure diagram

According to the landfill leachate treatment process, and then carry out the test operation in theengineering project. In order to realize the automatic operation of landfill leachate treatment, the automation control is carried out by using the host computer and the lower computer programmable controller (PLC) and the field sensor instrument.The structure diagram of the automatic control system of landfill leachate is shown in figure 2.The leachate treatment plant adopts the SCADA control system. Through industrial communication networks, field devices, control stations, and central control rooms communicate with each other to realize data interaction, form a complete communication network, and implement an automatic control system that integrates monitoring and management[18].

The host computer configuration king realizes the communication with the lower computer PLC through host link. This automatic control system has 3 PLC control stations, collecting data in the reaction pool through various sensors such as: liquid level value, pH value, COD value, turbidity, BOD_5value, ammonia nitrogen concentration, etc. Each PLC station realizes the data exchange through the PC link agreement. The controller mainly completes the functions such as sending out operation instructions, data collection and real-time processing, and abnormal phenomenon alarm processing. The upper computer adopts kingview to realize data monitoring for each PLC control station, and real-time monitoring of the operation of process equipment of each control station, and controls the execution equipment by adjusting the control parameters of each process.

Because of the components of landfill leachate are complex and the water quality changes greatly, the traditional PID control method has poor control accuracy in the leachate treatment process, and the parameters are difficult to adjust online. Therefore, a PID control method based on adaptive strategy is proposed and designed. The structure diagram of the BP-PID control structure is shown in figure 3. By using the self-learning ability and approximation capability of BP neural network, the PID parameters are adjusted online according to the characteristics of water quality change of leachate. And the controller's control effect and anti-jamming ability are enhanced.

4. Control System

The influent water quality measured by various sensor instruments is shown in table 1 below:

Table 1. Influent pollutant concentration table

Pollutants	Concentration range($mg \cdot L^{-1}$)	Pollutants	Concentration range($mg \cdot L^{-1}$)	Pollutants	Concentration range ($mg \cdot L^{-1}$)
COD	6000-8000	Cl^-	2000-3000	Pb	0.1-0.3
BOD_5	5000-7000	$SO4^{2-}$	400-600	Mn	30-40
TS	4000-6000	Ca^{2+}	3000-4000	Zn	50-70
SS	2500-3000	Fe	1000-1500	TCr	2-3
NH_3-N	2000-3000	Mg	500-600	VFA	200-300
P	40-60	Cd	2-3		
PH	5-9	Cu	3-4		

We carry out experimental operation in the project.After a series of physicochemical treatments, the refractory organics in the leachate are effectively oxidized and degraded, and the COD, BOD_5, ammonia nitrogen, suspended solids, chroma, and some heavy metal ions in the leachate are removed. The final effluent quality of leachate reached the first-class emission standards. The main indicators of leachate discharge standards after treatment are: pH6-9, COD≤100mg/L, BOD_5≤30mg/L, SS≤30mg/L, NH_3-N≤25mg/L, TN≤40mg/L, TP≤3mg/L.

We collected the average values of COD, NH_3-N, BOD_5, SS influent concentration and effluent concentration during the daily operation of the landfill leachate treatment system, and monitored a total of 25 days. The experimental results are shown in figure 4, Figure 5, Figure 6, and Figure 7.The removal rate was calculated and then the three parameters were plotted as a graph. As shown in the following four figures, we can see from the figure that the effluent concentration of the four pollutants is relatively low and the removal rate reached 99%.

IOP Conf. Series: Earth and Environmental Science **182** (2018) 012002 doi:10.1088/1755-1315/182/1/012002

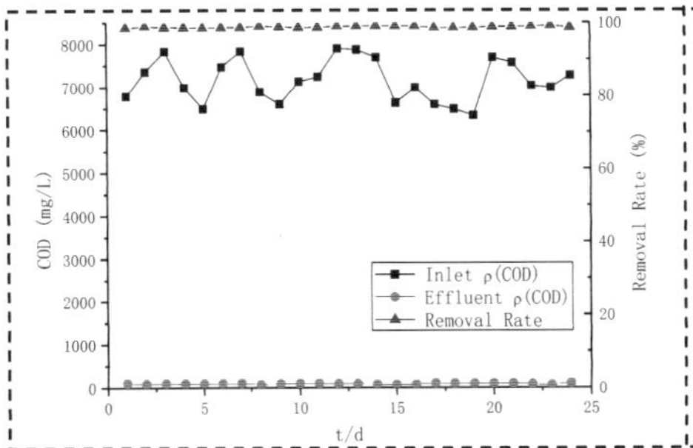

Figure4.COD removal effect

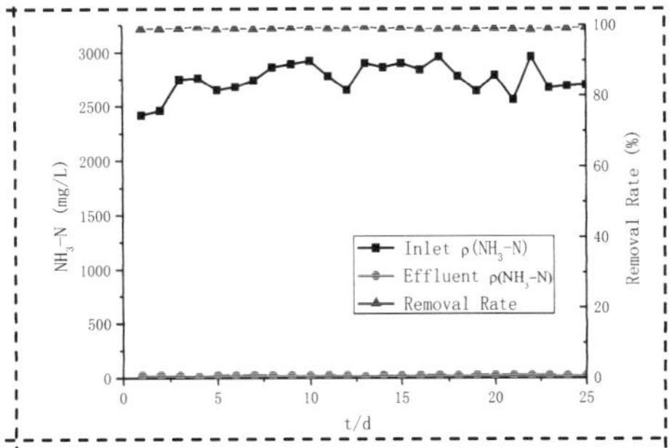

Figure5. NH_3-N removal effect

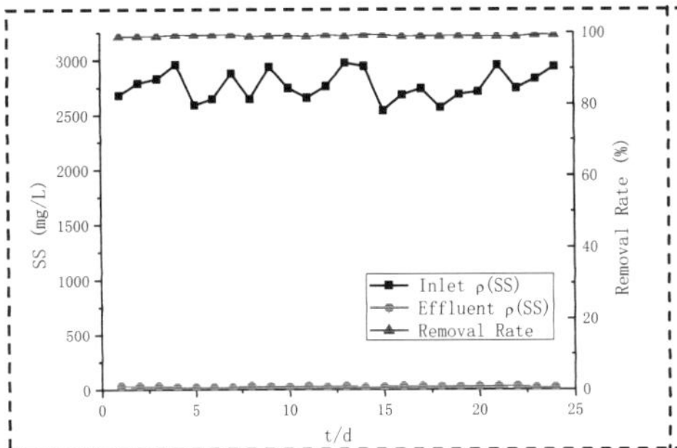

Figure6.SS removal effect

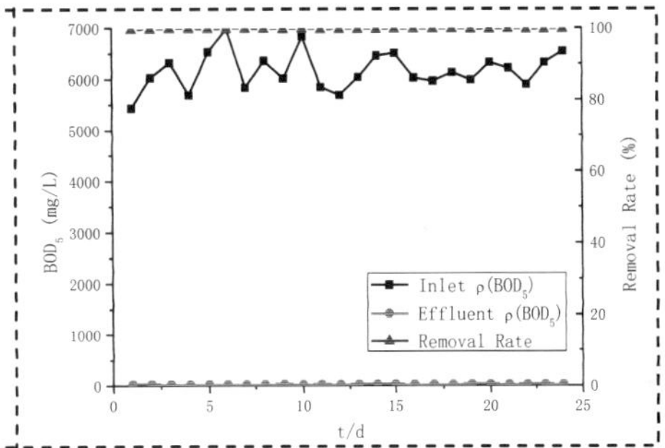

Figure7. COD removal effect

5. Conclusion

The water quality of landfill leachate is affected by external factors. The size of the landfill, the landfill operation mode, different seasons, and the local meteorological conditions all affect the quality of leachate. Leachate as a special high-concentration intractable wastewater, its investment and operating costs are much higher than the general urban domestic wastewater or industrial wastewater. The main reason is that the concentration of organic matter and ammonia nitrogen in the leachate is extremely high, the biochemical performance is poor, and the proportion of nutrients is unbalanced [2]. In this paper, physicochemical treatment is mainly applied to treat leachate. In general, the main advantages of the physicochemical method are stable treatment effect, assured water quality, and strong anti-hydraulic load capacity, especially with low biodegradability and low BOD_5/COD landfill leachate treatment [2].

This paper has achieved good results in the treatment of landfill leachate by ozone oxidation, internal electrolysis, peroxidation of persulfate, deep oxidation of chain reaction, microwave field catalytic oxidation, coagulation and precipitation, adsorption and other physical and chemical treatment. The process effectively oxidizes and degrades refractory organic matter in the leachate, removes COD, BOD_5, ammonia nitrogen, suspended solids, chroma, and some heavy metal ions in the leachate, and the effluent water quality meets the emission standards. At the same time, the process equipment has less investment, lower cost, short processing flow and high efficiency, and has a good application prospect in the treatment of landfill leachate. Through the application of automatic control system, the stability of the system has been improved, the precise control of the controlled equipment has been realized, the dosage of the medicine can be controlled more accurately, the treatment effect is better, the use of the medicine is reduced, the two pollution of the water quality is avoided and the cost of the sewage treatment is reduced.

However, there are also some defects in the treatment of landfill leachate, such as large amount of catalyst and low efficiency of some catalysts. Therefore, aiming at some problems in the leachate treatment process, we need to deeply study the mechanism of advanced oxidation technology reactions, strengthen the optimization and combination of advanced oxidation technologies and other treatment processes, and learn from each other to achieve high efficiency, cost-effective, and pollution-free treatment methods.Oxidants are the basis of physicochemical treatment methods. We must research and develop low-cost, more oxidation-efficient oxidants. The research and development of new and complex catalysts and the increase in the number of recycling catalysts will not only contribute to the treatment of landfill leachate, but also have important implications for the realization of their recycling and environmental protection. We also need to develop an efficient and simple reactor to treat landfill leachate, which will provide new ideas and directions for realizing landfill leachate treatment [3].

6. References

[1] XiaoC H2017Study on Leachate Treatment Process of Domestic Waste Landfill Site*Sci.Technol.Innov*111 112
[2] RenH Y, Han L, Kang Y L and Huang W W2016Process design and analysis of difficulties for treatment of landfill leachate *China Water Wastewater*32 56-58
[3] Liu Z B2017Research progress and analysis of treatment technology for landfill leachate*Guangzhou. Chem. Ind*32 38-41
[4] LI N, LI X M, YANG Qet al 2014Landfill leachate treatment bymicrowave-enhanced persulfate oxidation process using activatedcarbon as catalyst*China. Env. Sci*3491-96
[5] Amaral-silva N, Rui C M, Castro-silva S et al2016Ozonation and perozonation on thebiodegradabilityimprovementof a landfill leachate*J. Envion. Chem. Eng*4 527-533
[6] Tatsi A A, Zouboulis A I and Matis K A2003Coagulation-flocculationpretreatment of sanitary landfill leachates*Chemophere*53737-744
[7] Amor C, De T E, Peres J A et al2015Mature landfill leachatetreatmentbycoagulation/flocculation combined with fentonandsolar photo-fentonprocesses*J. Hazard. Mater*286 261-268
[8] Feng C J, Jiang L L, Liang T and Wu W H2015A ReviewontheGrilleEquipment in sewage treatment plant*Env. Eng*33 941-945
[9] Zeng S J2016Application of new grille decontamination machine in sewage treatment plant*Henan Building. Mat*5 88-90
[10] Wang W B, Dong Y and Liu S T2004Air stripping to remove ammonia nitrogen in landfill leachate*Techniques Equipment* 551-52
[11] Aosiman T, Yang L, An D and Wang T2014Progresses in air stripping for treatment of ammonia wastewater*Petrochemical. Techno*43 1348-49
[12] Liu Z B2018Domestic waste landfill site leachate deodorization process overview*Technology Equipment*44 86
[13] Liu Z M,Xu L C,Zhao J F et al2018Research progress of advanced oxidation treatment of landfill leachate*Technol. Water. Treat*447-11
[14] Amr S S A, Zakaria S N F and Aziz H A2016Performance ofcombined ozone and zirconium tetrachloride in stabilized landfillleachatetreatmentJ. Mater. Cycles Waste12 1-7
[15] Yang E L,Zhou Y X,Wang H L et al2016Experiment on the treatment effect of PAC-PAM treating the electroplating wastewater in one industrial park*J. Green. Sci. Technolo*16 141-143
[16] Zhu L S2010Study on the Degradation of Dyestuff Wastewater by the PAC/PAM FlocculationTreatment*Guangdong. Chem. Ind*37 123-126
[17] Mo D Q,Xiao W X and Chen B2007Adsorption filtration properties ofmodifiedquartz sand*J. Guilin. U. Technolog* 27 378-381
[18] Yang W X2018Critical review of landfill leachate treatmenttechnologies*Low. Carbon. World*23 59

Acknowledgments

Authors wishing to acknowledge Authors wishing to acknowledge HuiBaicompanyfor the data providing.

IOP Conf. Series: Earth and Environmental Science **182** (2018) 012003 doi:10.1088/1755-1315/182/1/012003

A Health Impact Assessment of Traffic Restrictions during Madrid NO_2 Episode

R. San Jose[1], J.L. Pérez[1], L. Pérez[1] and R.M. González[2]

1 Environmental Software and Modelling Group, Computer Science School, Technical University of Madrid (UPM), Madrid, Spain.
2 Department of Physics and Meteorology, Complutense University of Madrid (UCM), Madrid, Spain
E-mail: roberto@fi.upm.es

Abstract. One of the first decisions to improve the urban air quality during an air pollution episode is to apply traffic parking and access restrictions to try to decrease the amount of private vehicles driving in the city but their the effectiveness of the decisions must be evaluated before taking them. The health impact assessment tool of this work can help to the decision makers because it examines the citizen's health impacts of the applied measurements. The modelling system has been applied for a NO2 episode in Madrid city during December, 2016. The core of the system is the EMIMO-WRF/Chem air quality modeling system that simulates the air quality concentrations every grid cell of 1 km by 1 km and traffic emissions are calculated using data from a microscopic traffic model. The pollutant concentrations are inputs to the health impact module, which uses concentration–response functions. Two simulations were designed: "REAL" including traffic restrictions and "BAU" representing what would happen if no action were taken. The differences between the two simulations (BAU-REAL) give us the contribution of traffic restriction measures to improve the citizen´s health. The results show that the measures taken in this specific case were not sufficiently effective compared to the effort to reduce traffic.

1. Introduction

In the last years the link between air pollution and the health of citizens has been increasing in relevance in society and policy makers have started to develop strategies to try to minimise the health impacts on the health of the population [1]. Epidemiological studies have evidenced relations between daily concentrations of outdoor air pollution to adverse health effects. Air quality is linked to rising mortality and morbidity in European cities. In the last few years, various epidemiological studies have shown associations between an increase in daily concentrations of ozone (O_3), particulate matter (PM) and nitrogen dioxide (NO_2) and an elevation in mortality and/or hospital admissions in the following days, which are predominantly caused by respiratory and cardiovascular problems. These short-term health effects have been well documented in multi-centre time series studies [2] [3] [4] [5]. Also there are health impact assessment studies which have reported on the long-term effects of air pollution [6].Traffic is one of the most important sources of emissions in the cities, so recent studies try to investigate the scientific evidence for linking traffic emissions to negative health effects.
The levels of pollution in a city involve sophisticated physical and chemical processes that affect both formation and atmospheric transport. All these atmospheric processes are modeled by chemical transport models can be used to know the expected air pollution concentrations and to estimate health impacts of policies. This type of air quality models takes into account the emissions, chemistry and

transport of the pollutants. Health impact assessment (HIA) is a methodology for quantifying and evaluating the impact of air pollution mitigation efforts on human health [7, 8]. Mortality and morbidity are the most significant criteria for measuring the impact on the health of the people exposed to air pollution [9]. The general principle for a health impact assessment is to use information on how a change on air pollution concentrations is expected to modify the risk of disease or death of the citizens. The relation between exposure to pollutants and specific health outcomes is supported by the consistency of epidemiological findings across different studies. The most often used exposure indicators in HIAs has been particulate matter (PM) and nitrogen dioxide (NO_2) mass concentration for the effects of short-term exposure on mortality [10].

The methodology proposed in this work allows knowing a priori the benefits in the health of the citizens that would be expected after applying measures to reduce the exposure of the population to air pollution, such as traffic restriction measures in a city. The effectiveness of the measures to be taken can be evaluated before applying them and decide which may be the best solution for the health of citizens. The information produced by this kind of tools help managers and policy agents better assess the impact of theirs interventions.

The main objective of this study is to estimate the expected health benefits associated with traffic restrictions that produces mobility problems for the citizens. The work focuses on to investigate associations between short-term exposure to air pollutants daily mortality and morbidity in Madrid (Spain) during an important NOx episode on December 2016. The paper presents an integrated air quality modelling and health impact assessment tool, which can be used to evaluate the effectivity of decisions to reduce pollutant concentrations from a health point of view. Madrid is a city of about 3.5 million inhabitants with a population density of 5208 inhabitants/km2. The city is surrounded by 4 ring roads, the M30 being the limit of the Central District. The road network in the city centre is dense with fairly high traffic volumes. In 2016, the city of Madrid approved a new protocol for high levels of nitrogen dioxide pollution.

Four scenarios are considered depending on the pollution concentration of the different measuring stations within the city. The scenarios added new traffic restriction measures as the alert level increased. Traffic restriction measures range from a reduction of the speed limit on the M30 and road access to the city to a complete traffic restriction in the city centre. Intermediate scenarios also consider a downtown parking restriction and a partial traffic restriction depending on the license plate.

2. Material and Methods

In this section we described the air pollution episode and the air quality modelling system which was run to simulate the describe episode. Results from the air quality modelling system has used by the health impact assessment module.

2.1. Episode

We have selected a 5 days long NOx episode in Madrid. It was a very interesting episode because it was the first time and unique at the moment that the Madrid city apply a restriction of access to the city centre for private vehicles in order to reduce air pollution. In December 2016, the levels of $N0_2$ in Madrid were so high that authorities restricted access to the city centre for half of the cars based on whether the number plate was even or odd. The episode occurred from December 26 to 30, 2016, during which NO2 hourly concentrations reached 200 $\mu g/m^3$ in several monitoring stations. On Wednesday, December 28, the city temporarily banned parking in the city center by non-resident car owners and restricted speed limits on the main highway (M30) to 70 km/h instead of 90 km/h. Non-residents were prohibited from parking from 9:00 a.m. local time until 9:00 p.m. within the regulated parking areas.

The restriction of access to the city centre for private vehicles was applied on Thursday 29th December, only the odd number plates could access the inner area delimited by the M30 road (city centre). It was activated between 6:30 a.m. and 9:00 p.m. The measure was activated when the previous day's nitrogen dioxide levels in the atmosphere exceeded 180 $\mu g/m^3$ and predictions did not predict the weather conditions needed to improve air quality the next day.

Two scenarios were developed to examine associations between traffic restrictions (changes of traffic volume) and pollutant concentrations expressed as health impacts (changes on mortality and morbidity). The first scenario has considered the real traffic situation of those days that included traffic restrictions on Wednesday and Thursday, this simulation has been called "REAL" and includes traffic restrictions on Wednesday 28 (parking) and Thursday 29 (access). In the second scenario, we have deactivated the restrictions on Wednesday and Thursday, considering that those two days traffic followed a pattern similar to Wednesday and Thursday of the previous week in which there were no restrictions. Therefore, in this simulation, the same configuration and input data has been maintained except that the traffic emissions are different considering a "typical" Wednesday and Thursday traffic day, so this simulation has been called "BAU" (Business As Usual). Then the only difference between the "REAL" and "BAU" simulation is that there are no traffic restrictions applied in the BAU. The BAU simulation represents what would have happened if no traffic restriction measures had been taken on Wednesday and Thursday. The difference between the two simulations (BAU-REAL), gives us the contribution of traffic restriction measures to reduce concentrations of pollutants in the city of Madrid. This contribution may be either positive (traffic restrictions have reduced concentrations) or negative (restrictions have not improved air quality, but have aggravated pollution by increasing concentrations relative to BAU simulation).

2.2. Air quality modelling system

The EMIMO-WRF/Chem air quality modeling system has been uses for calculating the emissions and concentrations for the Madrid area with 1 km of spatial resolution. We have run air quality simulations using the Weather Research and Forecasting and Chem model with version 3.8.1 (WRF-Chem) [11] to study the NO2 episode in Madrid with a spatial resolution of 1 km by 1 km. The Carbon Bonding Mechanism version Z (CMBZ) is the atmospheric chemical mechanism [12] used for gas phase chemistry. Aerosol chemistry is represented by the Model for Simulating Interactions and Aerosol Chemistry (MOSAIC) [13]. Dry aerosol deposition is simulated following the approach [14] and the wet deposition approach follows [15] and [16]. Photolysis rates are obtained from the photolysis scheme in Fast-J [17]. We include aerosol-radiant feedback in our simulation. The Rapid Radiative Transfer Model (RRTM) scheme [18] is used to represent both short-wave and long-wave radiation. We use [19] and the parameterization of the Grell-3d cumulus set [20]. This configuration was tested in phase 2 of the International Air Quality Assessment Model Assessment Initiative (AQMEII) [21]. The initial and lateral boundary conditions of the meteorological variables, every six hours, were taken from the 0.5^{o} grid data of the Global Forecasting System (GFS) operated by the National Meteorological Service of the United States (NWS). The chemical conditions of the lateral boundaries for mother domain were taken from profiles.

TNO-MACC-II emission inventory [22] was processed using the EMIMO emission model [23]. Emission were speciated [24] and spatially and temporally to 1 km grid taking into account surrogate files. Biogenic emissions were calculated from the Guenther online scheme in the model [25]. Emissions are calculated for all important sectors but the difference in emissions in the two analyzed scenarios (BAU and REAL) are only due to differences in traffic emission because the traffic activity is different between the two scenarios. Traffic emissions were modeled taking into account changes in traffic situation of the two traffic scenarios (BAU, REAL). Traffic emissions were carried out using the Tier 3 method described in the EMEP/EEA 2016 Air Pollutant Emissions Inventory Guide - Update Dec. 2016 (Passenger cars, light commercial trucks, heavy vehicles including buses and motorcycles) that include specific emission factors and cover different engine conditions, following the COPERT methodology. [26]. Vehicle categories are divided by fuel type, vehicle weight, age of the vehicle and cubic engine capacity, each of which has its specific emission factors, defined according to traffic speed.

2.2.1. Traffic model

The traffic intensities and vehicle speeds are calculated by a traffic model. The microscopic SUMO traffic simulation [27] can be used to determine the large-scale effects of traffic management measures. The first entry to SUMO is the road network, which describes the part of a map related to traffic, roads and intersections that simulated vehicles travel. The road network has been obtained from OpenStreetMap. The road network consists of more than 100,000 streets and road segments. After you have generated a network, the next step is to put the vehicles in the network. SUMO allows you to use traffic detector data to generate traffic demand.

The information collected from traffic sensors can be used to construct vehicle quantities and routes. First, random traffic is generated for the Madrid network and then the detectors have been used as calibrators, which have been used to adapt traffic demand to a certain set of measures. Hundreds of traffic simulations have been carried out for each day, with different route configurations and the best of each day has been chosen. For the calibration of the SUMO model, data from more than 3,000 traffic counters were available, of which 2/3 were used for calibration and 1/3 for evaluation.

Traffic simulation underestimated traffic flow (vehicles/hour) by 7.8%. The adjustment obtained in the model calibration process shows a good convergence between the actual traffic flow data and the results obtained with the corresponding R^2 of 0.97. The composition of the fleet was collected from vehicle registration information in Madrid for December 2016. In addition to the vehicle and fuel type, classification also takes into account the vehicle's engine type, vehicle technology (age of vehicles). More than 600 vehicle categories power the emissions model. In December 2016, Madrid had 43,97972 vehicles registered, depending on the composition of the Madrid car fleet, more than half of the vehicles circulating with diesel (57.74%) and 27.06% of the total number of vehicles in Madrid are more than 15 years old. Diesel vehicles and their ages are two important factors in air pollution problems. Most vehicles are passenger cars (78.34%) with about 10% of motorcycles and motorcycles.

2.3. Health impact assesment

Using the health impact assessment module we calculate the estimated change in human mortality and morbidity between BAU scenario and the REAL emission scenarios for each of the 1 km grid cells. For our analysis of health impacts and potential benefits, we base on the US EPA's Benefits Mapping and Analysis Program (BenMAP) [28]. Our inputs to health impact module include modeled pollutant concentrations (daily averaged concentrations of NO_2 and $PM_{2.5}$) and the linkages between daily concentrations of some air pollutants and the risk of harmful effects on human health. The relationship between exposure variables and their effects on health can be modelled using log-linear (Poisson) regression and this function is called the exposure-response (ER) function. If we derive this function we get the equation (1) that allows us to estimate the change in mortality or morbidity as a result of a change in the respective exposure variable.

$$\Delta y = y_0(e^{\beta\Delta C} - 1) \quad (1)$$

where y_0 is the baseline incidence rate of the studied health effect, β is a parameter that gives us an estimate of the effect of mortality and that has been obtained from epidemiological studies, ΔC is the change of the exposure variable (BASE minus REAL) [29].

The concentration-response functions used by the health impact module are published in international epidemiological studies of high scientific acceptability. Changes in concentrations are used as input to the log-linear relationship between changes in concentrations and changes in mortality or morbidity. These are the relationships between the mean/maximum daily concentrations of air pollutants and the risk of hazardous health effects on the same day or on days after exposure at these levels, also taking into account the meteorological conditions on those days. The concentration-response functions used in our tool are based on the relative risks (RR) found in scientific studies that have found correlations between decreased concentrations of $PM_{2.5}$ and NO_2 and health benefits. They provide relationships between mortality or hospital admissions from cardiovascular and respiratory causes and levels of exposure to air pollutants. The used RR values are recommended by the HRAPIE project (Recommendations for concentration-response functions for cost-benefit analysis of particulate matter, ozone and nitrogen dioxide).

For the mortality and morbidity analysis the following exposure-response (E-R) relationships from studies were used: mortality all causes, RR=1.0027 (95% CI 1.0016 - 1.0038) for 10 μgm^{-3} increase of daily maximum NO_2 concentration; hospital admissions respiratory diseases, RR=1.0015 (95% CI 0.9992 - 1.0038) for 10 μgm^{-3} increase of daily maximum NO_2 concentration; mortality all causes, RR=1.0123 (95% CI 1.0045 - 1.0201) for 10 μgm^{-3} increase of daily mean $PM_{2.5}$ concentration; hospital admissions cardiovascular diseases, RR=1.0091 (95% CI 1.0017 - 1.0166) for 10 μgm^{-3} increase of daily mean PM2.5 concentration; hospital admissions respiratory diseases, RR=1.0190 (95% CI 0.9982 - 1.0402) for 10 μgm^{-3} increase of daily mean PM2.5 concentration. The concentration-response functions have associated an uncertainty so that in epidemiological studies the data are published with a 95% confidence interval.

3. Results

The text of your paper should be formatted as follows. Traffic simulations of SUMO REAL and BAU show that on day 28 (parking restrictions) traffic was reduced by 10.24%, on day 29 (access restrictions) traffic was reduced by 16%, 75% and on day 30 (parking restrictions) traffic was reduced by 6.06% If we focus on the city centre (within the M30) on day 28 the reduction only reached 4%, but on day 29 it reached 20%. It seems clear that access restriction measures were more effective in reducing traffic in the city centre, while parking restrictions for non-residents affected more vehicles from outside the centre of Madrid, as they could not park and used other modes of transport to arrive to the city. The main differences between REAL and BAU simulations can be observed on day 29, especially during the early hours of the day when people go to work. On the 30th, which was Friday parking restrictions did not reduce traffic almost in the afternoon, but on the 28th (Wednesday) the reduction is maintained throughout the whole day (morning and afternoon). If we now compare Madrid's emissions for these days, we can see that on day 28 parking restrictions reduce the emission of NOx to -8.27%, on day 29 -10.28% and on day 30 the reduction is only -1.77%.

In this study we analyse the health impact for the change in the daily concentrations of $PM_{2.5}$ and NO_2 between the two scenarios REAL and BAU. The impact of air pollution on mortality and morbidity was quantified in relative terms of the number of attributable deaths or hospital admissions. Figure 1 shows the calculated decrease in mortalities due to changes in NO_2 between the REAL scenario (traffic restrictions) and the base case BAU (without traffic restrictions) for the days 29 (traffic access restriction) and 30 (traffic parking restriction) of December, 2016.

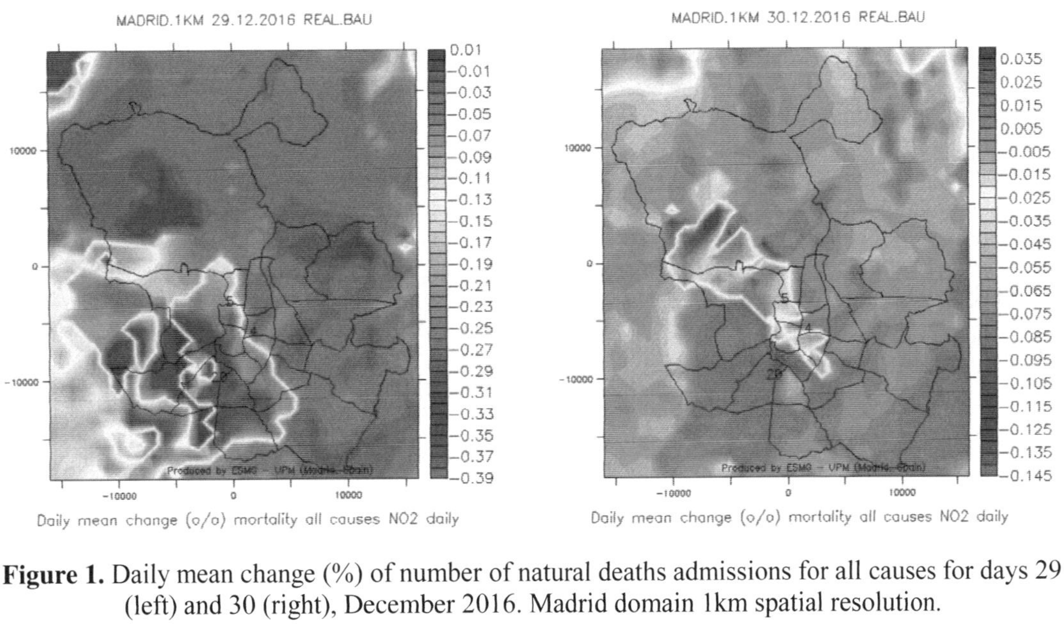

Figure 1. Daily mean change (%) of number of natural deaths admissions for all causes for days 29 (left) and 30 (right), December 2016. Madrid domain 1km spatial resolution.

Figure 1 shows important differences in the spatial distribution of the health impacts. On day 29, the health benefits of traffic access restrictions are located in the center and south-west part of Madrid city. On day 30, center and North West appear to experience the highest level of benefit. On both days the center is the area where traffic restrictions are applied because that this is where the air quality problems are most severe, but the health impacts are following the predominant wind directions (not showed). The highest health impact is found primarily in Madrid city center, where road transport density is highest. The mortality reduction is more important on day 29 (-0.39 %) that day 30 (-0.14 %). These reductions will result in 0.007 per 100,000 inhabitants fewer premature deaths on day 29 and 0.003 per 100,000 inhabitants on day 30 from traffic restrictions using NO2 as indicator.

4. Conclusions

This study demonstrates a methodology for analyzing the health impacts attributable to traffic restrictions. The modelling tool includes an emission model (EMIMO), which includes the SUMO model for traffic and a transport and pollutant chemistry model (WRF/Chem). The modelling system has been used to simulate an episode of high NO2 concentrations in the city of Madrid during December 2016 with high spatial resolution (1 km). The modeling system has been used to assess the effectiveness of the traffic restriction measures (parking restrictions at points 28 and 30; access limitations at point 29) taken by the Madrid City Council to try to reduce NO2 concentrations. The evaluation was carried out by comparing the REAL simulation (with traffic restrictions) with a BAU simulation (without traffic restrictions).

The results of the modelling of the estimated health impacts show that there would be insignificant reductions in mortality/morbidity due to the application of the proposed traffic restrictions. Therefore, the measures applied were not effective since the benefits for the health of the citizens were marginal. Using the health impact assessment methodology it is estimated that that less than one case of natural mortality will be avoided from the introduction of the traffic restrictions measurements. The expected change of NO2 by banning traffic from the Madrid city center is not associated with a considerable mortality or morbidity decrease. These results show that the measures taken were not sufficiently effective compared to the effort to reduce traffic. Other measures should be evaluated with less impact on citizens and with a greater capacity to reduce air pollution (transformation of diesel fuel into electric vehicles, prohibition of driving vehicles over 15 years old, reduction of traffic speed, etc.).

5. References

[1] Künzli N, Kaiser R, Medina S, et al. Public-health impact of outdoor and traffic-related air pollution: a European assessment, *Lancet*, 2000, vol. 356 9232(pg. 795-801).

[2] Anderson HR, Atkinson R, Peacock JL, Marston L, Konstantinou K. Meta-analysis of time series studies of particulate matter (PM) and ozone (O3). EUR/04/5042688. Copenhagen: *WHO Regional Office for Europe*; 2004. p. 1-80.

[3] Atkinson R, Anderson HR, Medina S, Iniguez C, Forsberg B, Segerstedt B, et al. Analysis of all-age respiratory hospital admissions and particulate air pollution within the APHEIS programme. *APHEIS health impimpact assessment of air pollution and communication strategy. Third year report*; 2005.

[4] Garrett P, Casimiro E. Short-term effect of fine particulate matter (PM(2.5)) and ozone on daily mortality in Lisbon, Portugal. *Environ Sci Pollut Res Int* 2011

[5] Ruckerl R, Schneider A, Breitner S, Cyrys J, Peters A. Health effects of particulate air pollution: A review of epidemiological evidence. *Inhal Toxicol* 2011; 23: 555-92

[6] Cohen AJ, Anderson HR, Ostro B, et al. Ezzati M, Lopez AD, Rodgers A, et al. Mortality impacts of urban air pollution, Comparative Quantification of Health Risks: Global and Regional Burden of Disease Due to Selected Major Risk Factors , 2004, vol. Vol 2 Geneva, SwitzerlandWorld Health Organization

[7] U.S. EPA (U.S. Environmental Protection Agency), 2010: BenMap: Environmental Benefits Mapping and Analysis Program User's Manual, Appendix, Research Triangle Park, NC:U.S. EPA, Office of Air Quality Planning and Standards.

[8] PE Sheffield, K. Knowlton, JL. Carr , PL. Kinne, " Modeling of regional climate change effects on ground-level ozone and childhood asthma," *Am J Prev Med.* 2011;41(3):251–257.

[9] Hong, S.Y., Pan, H.L., 1996: Nonlocal boundary layer vertical diffusion in a medium-range forecast model. *Monthly Weather Review*, 124(10), 2322–2339.

[10] WHO. Review of evidence on health aspects of air pollution – REVIHAAP project: final technical report. WHO, Bonn, 2013.

[11] Grell, G. A., Peckham, S. E., Schmitz, R., McKeen, S. A., Frost, G., Skamarock, W. C., and Eder, B.: Fully coupled "online" chemistry within the WRF model, *Atmos. Environ.*, 39, 6957–6975, doi:10.1016/j.atmosenv.2005.04.027, 2005.

[12] Zaveri, R. A. and Peters, L. K.: A new lumped structure photochemical mechanism for largescale applications, *J. Geophys. Res.*, 104, 30387–30415, 1999.

[13] Zaveri, R. A., Easter, R. C., Fast, J. D., and Peters, L. K.: Model for simulating aerosol interactions and chemistry (MOSAIC), *J. Geophys. Res.*, 113, D13204, doi:10.1029/2007JD008782, 2008.

[14] Binkowski, F. S. and Shankar, U.: The Regional Particulate Matter Model: 1. Model description and preliminary results, *J. Geophys. Res.-Atmos.*, 100, 26191–26209, doi:10.1029/95JD02093, 1995.

[15] Easter, R. C., Ghan, S. J., Zhang, Y., Saylor, R. D., Chapman, E. G., Laulainen, N. S., Abdul-Razzak, H., Leung, L. R., Bian, X., and Zaveri, R. A.: MIRAGE: Model description and evaluation of aerosols and trace gases, *J. Geophys. Res.-Atmos.*, 109, D20210, doi:10.1029/2004JD004571, 2004.

[16] Chapman, E. G., Gustafson Jr., W. I., Easter, R. C., Barnard, J. C., Ghan, S. J., Pekour, M. S., and Fast, J. D.: Coupling aerosolcloud- radiative processes in the WRF-Chem model: Investigating the radiative impact of elevated point sources, *Atmos. Chem. Phys.*, 9, 945–964, doi:10.5194/acp-9-945-2009, 2009.

[17] Wild, O., Zhu, X., and Prather, M.: Fast-J: Accurate Simulation of In- and Below-Cloud Photolysis in Tropospheric Chemical Models, *J. Atmos. Chem.*, 37, 245–282, doi:10.1023/A:1006415919030, 2000.

[18] Mlawer, E. J., Taubman, S. J., Brown, P. D., Iacono, M. J.,and Clough, S. A.: Radiative transfer for inhomogeneous atmospheres: RRTM, a validated correlated-k model for the longwave, J. Geophys. *Res.-Atmos.*, 102, 16663–16682, doi:10.1029/97JD00237, 1997.

[19] Lin, Y.-L., Farley, R. D., and Orville, H. D.: Bulk Parameterization of the Snow Field in a Cloud Model, *Journal of Climate and Applied Meteorology*, 22, 1065–1092, doi:10.1175/1520-0450, 1983

[20] Grell, G. A. and Dévényi, D.: A generalized approach to parameterizing convection combining ensemble and data assimilation techniques, *Geophys. Res. Lett.*, 29, 38-1–38-4, doi:10.1029/2002GL015311, 2002

[21] San Josè, R., Pèrez, J. L., Balzarini, A., Barò, R., Curci, G., Forkel, R., Galmarini, S., Grell, G., Hirtl, M., Honzak, L., Im, U., Jimènez-Guerrero, P., Langer, M., Pirovano, G., Tuccella, P., Werhahn, J., and Zabkar, R.: Sensitivity of feedback effects in CBMZ/MOSAIC chemical mechanism, *Atmos. Environ.*, 115, 646–656, doi:10.1016/j.atmosenv.2015.04.030, 2015

[22] Kuenen, J. J. P., Visschedijk, A. J. H., Jozwicka, M., and Denier van der Gon, H. A. C.: TNO-MACC_II emission inventory; a multi-year (2003–2009) consistent high-resolution European emission inventory for air quality modelling, *Atmos. Chem. Phys.*, 14, 10963-10976, https://doi.org/10.5194/acp-14-10963-2014, 2014

[23] San Jose R, Juan L. Perez, Jose L. Morant, Rosa M. Gonzalez, European operational air quality forecasting system by using MM5-CMAQ-EMIMO tool, *Simulation Modelling Practice and Theory*, Volume 16, Issue 10, 2007

[24] Tuccella, P., Curci, G., Visconti, G., Bessagnet, B., Menut, L., and Park, R. J. (2012), Modeling of gas and aerosol with WRF-Chem over Europe : evaluation and sensitivity study, *J. Geophys. Res.*, 117.

[25] Guenther, A.B., Zimmerman, P.R., Harley, P.C., Monson, R.K., Fall, R., 1993. Isoprene and monoterpene emission rate variability: model evaluations and sensitivity analyses. J. Geophys. Res. 98, D7. http://dx.doi.org/10.1029/93JD00527

[26] Markakis, K., Valari, M., Colette, A., Sanchez, O., Perrussel, O., Honore, C., Vautard, R., Klimont, Z., and Rao, S., 2014: Air quality in the mid-21st century for the city of Paris under two climate scenarios; from the regional to local scale. *Atmos. Chem. Phys.*, 14, 7323-7340

[27] Krajzewicz, D., J. Erdmann, M. Behrisch, and L. Bieker (2012) Recent Development and Applications of SUMO - Simulation of Urban MObility. In: *International Journal On Advances in Systems and Measurements*, 5 (3&4), pp. 128-138. ISSN 1942-261x.

[28] Abt Associates Inc: BenMAP, Environmental Benefits Mapping and Analysis Program, User's Manual, Prepared for the US EPA Office of Air Quality Planning and Standards, available at: http://www.epa.gov/airquality/benmap/docs.html August 2010

[29] U.S. EPA (U.S. Environmental Protection Agency), 2010: BenMap: Environmental Benefits Mapping and Analysis Program User's Manual, Appendix, Research Triangle Park, NC:U.S. EPA, Office of Air Quality Planning and Standards.

The Analysis of Control Technology of Non-point Pollution and Endogenous Pollution of the Water Environment

Xin Zhang[1], Peng Li[2], Xigang Xing[3], Zhuoran Wang[4], Xijun Gong[5]

1.2.3. Zhongke Dingshi Environmental Engineering co. LTD
4. Yanbian University
5. North Control Water (China) Investment co. LTD
E-mail: 1035852509@qq.com

Abstract: With the development of globalization, regional integration, industrialization and urbanization, the problem of water environment treatment in China has become more complex, more serious and contradictory. In order to solve the water environment problem fundamentally, the source of pollution should be cut off from the source. Non-point pollution has become one of the main sources of pollution in the basin, so blocking and treating the non-point pollution is also a test for environmental practitioners. The paper introduces the treatment technology of the non-point source pollution and the endogenous pollution from many aspects. At the same time, the paper also expounds the condition of usage and the range of application.

1. The technology of non-point source pollution treatment

1.1. The source of non-point source pollution

The main sources of non-point source pollution include the rural non-point source pollution, the runoff pollution and the dry and wet deposition pollution. The rural non-point source pollution is made up of the aquaculture pollution, the planting pollution and the rural domestic sewage. The runoff pollution is made up of the initial rainwater in the rainwater pipe network and the initial runoff in the non-rainwater control area. It also includes the dry and wet deposition pollution which is entering into rivers directly.

1.2. The integrated control technology of the non-point source .pollution

The section introduces the integrated control technology of the aquaculture pollution, the planting pollution, the rural domestic sewage and the runoff pollution.[1]

1.2.1. The integrated control technology of the aquaculture pollution. The ultimate goal of the disposal of the aquaculture pollution is to realize the efficient use and recycling of resources and the minimization of the pollution. It is usually possible to separate solids from liquids. The separated waste water is used to generate electricity and supply heating by adopting the technique of efficient anaerobic biogas production technology. The other part is treated by the high efficiency purification process. The solid aquaculture pollution can be realized ecological utilization within the range of region. The aquaculture pollution emission is strictly controlled from the source through the above integrated management measures.

1.2.2. The integrated control technology of the planting *pollution.* In order to solve the problems of excessive fertilizer application, low level of management of water fertilizer and high nitrogen and phosphorus runoff, the measures of environmentally friendly plastic, water-saving irrigation, moderate fertilization and nitrogen and phosphorus ecological interception can be used to control the disorder of non-point source pollution emissions from the source, the erosion process and the end of the link.
Take the example of rice production, applying rice slow-release fertilizer can effectively control the nutrient release speed, extend the fertilizer effect which fully meet the needs of the whole growth period and practically match the growth line, meanwhile, which can avoid the excessive application of fertilizer and improve the utilization rate of fertilizer. Besides, some simple tools such as the rice leaf ratio color card and the field water level monitoring cylinder can be used to achieve accurate fertilization and moderate irrigation and reduce the risk of nitrogen and phosphorus loss. In the end, the ecological ditch or ecological detention pond can be constructed depending on the yield of water to intercept nitrogen and phosphorus terminally in the field of returning farmland. The current common treatment methods include centralized sewage treatment and decentralized sewage treatment.

1.2.3. The integrated control technology of the rural domestic sewage. The rural domestic sewage has the characteristics of good dispersion and difficult centralized collection processing, which has become an important river pollution source [2].
Take the example of coupling treatment technology of anaerobic + multi-medium soil. Sewage goes through the septic-tank, the anaerobic bank and the artificial reinforcement multi-media soil composed of permeable layer in turn to achieve purification. The treatment technology has advantages of high hydraulic loading, excellent performance of nitrogen and phosphorus removal, low cost, free energy and strong adaptability, which has been successfully applied.

1.2.4. The integrated control technology of the runoff pollution. The early runoff pollution is high concentration and heavy load, which has a great influence on the quality of the receiving water[3]. The main measures which are used to control the runoff pollution include rainwater storage tank, high efficiency hydraulic cyclone separation device, low impact development design (LID) and ecological buffer zone and so on.
For example, the initial rainwater can be intercepted and stored by the storage tank. And when it's sunny, the initial rainwater can be transported to the sewage plant by submersible sewage pumps for treatment
The above measures can greatly reduce the discharge of initial rainwater into the river. Efficient hydraulic cyclone separation technology is based on the principle of physical separation, which can effectively separate the particles, scum and grease from rainwater runoff.. Low impact development (LID) and ecological buffer technology which are using the principle of ecological interception greatly enhance the interception effect of non-point source pollution, thus reducing runoff and improve the quality of receiving waters.

1.3. The ecological restoration of non-point source pollution
Carrying out the ecological restoration of polluted water and biodiversity restoration work is the main study direction and important engineering measures of non-point source pollution[4]. The water quality improvement and ecological restoration of polluted rivers are important measures to ensure the sustainable development of aquatic ecosystem.
According to the implementation position difference of engineering measures, the technology of water quality improvement can be divided into three categories: in situ repair, semi-in-situ repair, and heterotopic repair. In situ restoration is to set up some artificial oxygenation measures, artificial water grass and ecological gravel bed in the natural water to increase the oxygen content of water and provide sufficient habitat for aquatic organisms. Thus it can enhance the purification effect. The semi-in-situ restoration which is aimed at the river channel with smooth both or three sides is to upgrade the polluted water to the top of the revetment, then go through the high efficient purification medium laying on the bank to purify the water. Heterotopic restoration is a process that transfer the river water to other locations on the shore and discharge treated water into the river. The common

heterotopic restoration technology includes air flotation decontamination technology, super magnetic separation decontamination technology and bypass intensified wetland technology and so on.

1.3.1. The air flotation decontamination technology. The air flotation decontamination technology is a solid-liquid separation process. The water purification is a process that the hydrophobic impurities in the water collide with the water hydrophobic micro-bubbles under certain hydraulic conditions. Then the two stick together through the VDW and float to the surface of the water and remove from the water.
The air flotation decontamination devices mainly include the container system, the release gas system, the separation system and the slag discharge system. The working principle of the air flotation decontamination devices is shown as below: The air is attached to the suspended particles in highly dispersed microbubbles. The result is that the density of the suspended particles is less than the water. Then the suspended particles float on the water surface and realize solid-liquid separation by buoyant effect. The air flotation methods include bulk air flotation, dissolved air flotation and electrolytic air flotation. Compared with other solid-liquid separation devices, the air flotation decontamination devices have the advantages of low investment, small area, high automation and convenient operation management.

1.3.2. The super magnetic separation decontamination technology. The premise of the method of using the super magnetic separation decontamination technology is that the particles in sewage have certain magnetism. For the sewage with non-magnetic or weak magnetic, combing the nonmagnetic substance with magnetic seeds which are artificial additive. And purify the sewage by the super magnetic separation decontamination technology or flocculation settlement with high gradient magnetic separation decontamination technology. The technology has the advantage of good water quality, convenient operation management and low cost.

1.3.3. The bypass intensified wetland technology. The bypass intensified wetland system is a sewage treatment technology which is simulating natural wetland to achieve sewage treatment. The system is made up of packed beds, plants, aquatic animals and microorganisms.
It is widely used due to its strong processing ability, high efficiency and convenient management. The working process of the technology is to achieve to remove impurity especially organic pollutant by filtration, adsorption, precipitation, ion exchange, plant absorption and microbial decomposition.

2. The technology of endogenous pollution treatment

Endogenous pollution mainly refers to the nutriment settle down to the bottom of the lake gradually through all kinds of physical functions, chemical functions and biological functions. One the one hand, the nitrogen, phosphorus and other nutriment accumulated on the surface of the sediment can be eaten by microorganisms, then enter the food chain and participate in the circulation of aquatic ecosystems. On the other hand, the nutriment can release from the sediment and enter into the water under certain physical and chemistry conditions to become endogenous pollution.
The control and elimination of endogenous pollution is the most direct and effective way to cut down the pollution source of black and odorous water. The following is an introduction to the three control technology of endogenous pollution such as sediment pollution, aquaculture pollution and shipping pollution.

2.1. The technology of sediment pollution treatment

The study says, the amount of pollutant released by sediment is equivalent to that of external source. In addition, the algae blooms due to the large amount of nutriment in the urban river. These algae supply oxygen to the water in the early stage of growth, then resolve and mineralize to COD and NH_3-N after death, and it causes the phenomenon of the black and odorous water and produces an extremely strong odor.
The treatment techniques of sediment pollution include engineering treatment, chemical treatment, bio-ecological restoration treatment and resource utilization. The techniques of engineering treatment

include environmental dredging, mechanical dredging and solidification and stabilization. The techniques of chemical treatment include improver and inhibitor. The techniques of bio-ecological restoration treatment include plant restoration, microbial restoration and animal restoration. The resource utilization refers to land utilization, producing construction material and fill material.

2.1.1. The technology of engineering treatment

1. The environmental dredging
The environmental dredging is a method of removing endogenous pollution by combining with ecological restoration. It is achieved through environmental reamer. Dredger ships with high positioning accuracy and mining precision require that different sites have different depths and that the sediment pollution should be removed in a controlled way.
The environmental dredging machine should have the following characteristics: 1. Dredger should be with high accuracy and mining precision and avoid leakage and over excavation, and not harm the original soil. 2. Prevent disturbance and diffusion and don't cause the secondary pollution during digging. 3. Reduce the amount of water treatment by high concentration inhalation and as low as possible overflow.

2. The mechanical dredging
The mechanical dredging includes dredging rivers and lakes, dredging inland waterway and excavating lakes, channels and seaside. Its work includes dredging and filling. The filling is to carry mud to the operation surface with press pipeline to achieve the effect of dredging and strengthening.
The dredging ship is divided into mechanical, hydraulic and pneumatic three categories according to the work principle and delivery way. The main types includes grab dredger, dipper dredger, wheel dredger, pump-suction dredger, jet dredger and drag-suction dredger, in which cutter-suction dredger is the most widely used.

3. The solidification and stabilization
Dredging sediment is a kind of engineering waste soil which is high water content, small particles, high organic content, in the flow state and little intensity, and which can't be used for the project directly. The solidification and stabilization treatment is an effective resource utilization way which can turn the large amount of sediment into good engineering soil [5].. The core of the treatment is the selection and deployment of sludge stabilizer and modifier, and to realize the fast and efficient mixing reaction of sludge and modified agent. The water in the sludge is in the form of interstitial water and cellular water. The interstitial water is easy to be mixed with the modified agent. The cellular water reacts with agent after complete broken cells and micelles. The solidification and stabilization has the advantages of short period, fast effect, easy operation and low cost [6]..

2.1.2. The technology of chemical treatment
The method of chemical treatment mainly refers to add water and sediment quality modifier which is with physical or chemical activity to improve water quality and reduce the amount of harmful substances. The water and sediment quality modifier includes physical modifier, chemical modifier and micro-ecological agent.
Physical water quality modifiers generally have large specific surface area and many gaps. Poisonous and harmful substances are adsorbed to the sponge pore, so that ammonia, hydrogen sulfide and nitrite are reduced. In aquatic production, frequently used ones are zeolite powder, medical stone, activated carbon, alum and so on.
Chemical active water quality modifier are classified into oxidants, ion exchangers, heavy metal ions complexing agents, flocculants and so on. The principle of action is to oxidize ions in water, to exchange ions in water, and to degrade into compounds without secondary pollution. They are often specific and selective for other water quality substrates used in coordination.
At present, microbial ecological modifiers are the most important types of water quality modifiers [7]. They are able to break down organic ammonia in the water, such as excess organic matter, excreta,

feed, into harmless nitrogen compounds and nitrogen oxides. They can reduce the concentration of ammonia and nitrite, and promote the metabolism of anaerobic bacteria in the subsoil, such as sulfur bacteria, nitrifying bacteria and reducing bacteria. They can reduce toxic gases at the bottom of the pond, purify water and sediment, and restore microbial ecology of the aquaculture water. Commonly used microbial agents are EM bacteria, photosynthetic bacteria, Bacillus subtilis, nitrifying bacteria, denitrifying bacteria, yeast, Lactobacillus and other bacteria or mixtures[8].

2.1.3. The technology bio-ecological restoration treatment

The technology bio-ecological restoration treatment refers to reduce the concentration of poisonous and harmful substances in the environment or make it completely harmless by using the biological metabolism, so that the polluted environment can partially or fully recovered to the original state.

1. The plant restoration technology

The plant restoration technology is a kind of treatment method which is widely used in the field of environmental pollution. It reduces or eliminate the toxicity of pollutants by using the absorption, metabolism and elimination of the microorganisms of plant roots. There are three mechanisms of the plant restoration. The first is that plants absorb organic pollutants directly. The second is that the secretions and the enzyme released by plant roots resolve organic pollutants. The third is the combination of plants and root microorganisms.

A research by Anderson etc. [9] shows that plants can stimulate microorganisms to transform organic pollutants in many different ways, in which rhizosphere microorganisms play an important role. Because of plant roots, activity and quantity of microorganisms increased in soil. Microbial number and activity increased by 5~10 times in soil with plant roots some as high as 100 times, which will accelerate the degradation of many organic pesticides, trichloroethylene, petroleum hydrocarbon, methyl sulfur substances and some pesticides. Jordahl etc. [10] found that in the soil with Populus diltoides roots, the number of heterotrophic organisms, BTEX degrading microorganisms and herbicide Atrazine is higher than that without roots. Katayama etc. [11] also studied the degradation of various organic compounds such as pentachlorophenol, DDT with the fungi in the rhizosphere. They confirmed that fungi in the rhizosphere can play an important role in the degradation of organic pollutants probably because of microorganism activity is stimulated by secretions of roots.

Ma Weifang etc.[12] did an greenhouse experiment of , the specific surface area of the sediment was reduced by adding a certain amount of EDTA, citric acid and DTPA, during the Ryegrass remediation of heavy metals (Zn, Pb, Cu, Cd, Ni) - organic compound pollution of urban sewage river. It is beneficial for the release of heavy metals by sediment particles to add these, and also increases the number of microorganisms in rhizosphere and accelerates phytoremediation.

2. The microbial restoration technology

The microbial restoration technology is a kind of method that the organic pollutants are degraded into CO_2, H_2 and other harmless substances by oxygenation, reduction and hydrolysis of natural or domesticated microorganisms. The microorganisms can be made by artificial domestication, immobilized and GMO project.

Ma Weifang etc. [12] did an greenhouse experiment of , the specific surface area of the sediment was reduced by adding a certain amount of EDTA, citric acid and DTPA, during the Ryegrass remediation of heavy metals (Zn, Pb, Cu, Cd, Ni) - organic compound pollution of urban sewage river. It is beneficial for the release of heavy metals by sediment particles to add these, and also increases the number of microorganisms in rhizosphere and accelerates phytoremediation.

When Feng Qixiu etc. [13] are dealing with Chaoyang River in Guangzhou, without complete interception, they diluted and mixed 250kg culture solution of indigenous microorganisms (1 * 106pic/mL) and BE bio-stimulants 37.4L by the American company general with sewage dilution . Then the mixture was injected to the sediments through targeted delivery technique, in order to promote the oxidation. This process lasted for 5 days. Then 12.5L biological oxidation mixture every day was sprayed with the same method for 30 days. Then halve the dose and continually injected for 25 days. It is observed that dissolved oxygen content in water gradually increased both the upstream and downstream, and removal rates of COD, ammonia nitrogen and total phosphorus are also improved. Organic pollutants of aerobic microorganism system had been strengthened, and the purification ability of the overlying water significantly increased.

Duque etc. [14] transferred Pseudomonas patina plasmid with toluene degradation properties to Pseudomonas sp. Clong A, so it can oxidase TNT and use it as a nitrogen source for growth. It can acquire binding energy, degrade toluene and TNT, and cause TNT fully mineralized. Yan Yanchun etc. [15] cloned the esterase gene of Culex pipes pollens and expressed it in E. coli. The engineered bacteria were immobilized to degrade organophosphorus pesticides. The results showed that the bacteria degrading efficiency of pesticide was high in a short period.

Schippers etc. [16] studied the promotion of biological surfactant Sophorolipids on microbial degradation of phenanthrene. Two kinds of culture medium were used, including liquid medium and 10% soil suspension. The study showed that the highest degradation rate of phenanthrene increased significantly with the two additions of bio surfactants. Also, the concentration of residual phenanthrene decreased significantly. Further studies showed that this promotion is not because of the number of microorganisms increased, but the solubilization of phenanthrene by biological surfactants, which results in enhanced bioavailability of phenanthrene.

Wang Xin etc. [17] degraded phenanthrene and pyrene in soil pollutants with immobilization technique of Lactobacillus. The results showed that the immobilized microorganism had an absolute advantage in degradation of pollutants, since its effect is much better than autochthonous.

2.1.4. The resource utilization

The principles of the resource utilization of dredging sediment are harmlessness, reliability and stability. Choose the resource utilization method suit locally according to the sources, the composition characteristics of the dredging sediment and the local economic and technological conditions.

1. The fill materials

Make the dredging sediment suitable for the engineering requirement after improving the quality by pretreatment. Then backfill the treated sediment as fill materials. The methods of pretreatment include physical methods, chemical methods and heat treatments. The solidified soil has the advantages of no consolidation settlement, high strength and good permeability compared with general soil.

2. The land utilization

The land utilization is to apply dredging sediment to the restoration and reconstruction of the land which are farmland, woodland, grassland, wetland, municipal greening, seeding matrix and seriously disturbed land.

3. The producing construction

The dredging sediment can be used to make building materials, concrete lightweight aggregate and silicate gel materials. The building materials include ceramicite , bricks and tiles.

2.2. The technology of aquaculture pollution treatment

The technology of aquaculture pollution treatment includes: (1) Carry out the cleaning work and remove the net breeding gradually; (2) Implement the demonstration project of pond recirculating aquaculture; (3) Ensure the fishery environment and administer according to law.

2.3. The technology of shipping pollution treatment

The technology of shipping pollution treatment includes: (1) Strengthen anti-pollution legislation; (2) Reconstruction the suspended screw river boats and accelerate the standardization project of ship type; (3) Carry out the construction of stations which place the water waste and oil wastewater.

3. Reference

[1] Shuying He, Jixiang Li, Yatong Xu. Progress in the Research on the Management of Polluted Rivers [J]. *Journal of Henan Normal University (JCR Science Edition)*, 2008,36(2):75-78.

[2] Wei Meng, Yibing Su, Binghui Zheng. Discussion on the Present Situation and Control Strategy of Watershed Pollution in China [J]. *Journal of China Institute of water resources and Hydropower Research*, 2004,2(4):242-246.
[3] Kayhanian M, Singh A, Suverkropp C. Impact of annual average daily traffic on highway runoff pollutant concentration[J]. *Journal of environmental engineering*, 2003, 129(11):975-990.
[4] Wen Liu, Lihua Cui. The progress of the application of the artificial wetlands used in wastewater treatment[J]. *Journal of Jiayou University*, 2002, 20(3):29-32.
[5] Yiqun Gan, Yanxin Wang. Experimental study on Cu ion in wastewater with modified sludge removal[J]. *Environmental Science and Technology*, 2005, 06(2):223-225.
[6] Lei Li. Research on technology of sludge curing and control of heavy metal pollution[D]. *Hohai University*, 2006.
[7] Yongqing Huang, Xuehao Chen. The application of microecological in aquaculture[J]. *Feed Research*, 2004(7):42-43.
[8] Sekine K, Ohta J, OnishiM. Analysis of antitumor proper tiesol effector cells stimulated with a cellwall preparation of bifidobacterium infants. *Biolpharm. Bull*, 1995, 18(1):148-153.
[9] Anderson T A, Ellzabeth A G, Walton B T. Bloremedation in the rhizosphere[J]. *Environ Sci Technol*, 1993, 27(13): 2630-2635.
[10] Jordahl JI. Et al. Effect of hybrid poplar tree on microbial populations important to hazardous waste bioremediation[J]. *Environ Toxical Chem.* 1997, 16(6):1318-1321.
[11] Katayama A, Matsumura F. Degradation of organochlorine pesticides, particularly endosulfun, by Trichnoderma harzianum[J]. *Environ Toxical Chem*, 1993, 12:1059-1065.
[12] Weifang Ma, Xinhua Zhao, Jingmei Sun. The regulation effects of EDTA on plants restoring polluted dredging sediment[J]. *Environmental Science*, 2006, 7(1):85-90.
[13] Qixiu Feng, Jun Xie, Jun Liu. Sediment biological oxidation and city black and odorous water treatment[J]. *Reservoir Fisheries*, 2003, 23(6):42-44.
[14] Deque E. Construction of a Pseudomonas hybrid stain that mineralizes 2.4.6-trinitrotoluene[J]. *Bacteriol*, 1993, 175(8):2278-2283.
[15] Yanchun Yan, Liangtong Yao, Xiaoyan Song. The degradation of organophosphorus pesticides by engineered bacteria and immobilized cells[J]. *China Environmental Science*, 2001, 21(5):412-416.
[16] Schippers C. Gebner K. Muller T. Microbia degradation of phenanthrene by addition of a sophorolipid mixture[J]. *Biotechnol*, 2000, 83(3):189-198.
[17] Xin Wang, Peijun Li, Zongqiang Gong. Using the immobilization technique to deal with the contamination og pyrene in soil[J]. *Environmental Science*, 2002, 23(3):84-87.

Geographical Distribution and Risk Assessment of Heavy Metals in Nearby River of Heap Bioleaching Plant: A Case Study At the Zijin Copper Mine, China

MingJiang Zhang[1], MinJie Sun[2], XingYu Liu[1*], YiBin Li[1] and Xiao Yan[1]

1 National Engineering Laboratory of Biohydrometallurgy, GRIMAT Engineering Institute Co., Ltd. (GRIMAT), No. 11 XingKeDongda Str., Huairou District, Beijing 101407, China
2 China Certification & Accreditation Institute, A10 Chaowai Street, Beijing, China 100020

Abstract. The Zijin heap bioleaching plant was operated at the end of 2005. Concerns about the potential risk of environmental pollution from heap bioleaching plant arise due to the proximity to the Ting River. In this study, a physicochemical analysis, a geo-accumulation index and a high throughput sequencing technology were applied to determine heavy metals, assess the extent of heavy metal pollution, and research the effect of the heap bioleaching plant on the microbes, respectively. Results showed that the heap bioleaching plant had significant influence on the distribution of S, Pb and Cu and no significant influence on the distribution of As, Fe and Cr. Most of the water samples reached the third class standard of the People's Republic of China for surface water and individual water samples were above the fifth class standard of the People's Republic of China for surface water (GB3838-2002) because of As. The heap bioleaching plant had some effect on the microbial biomass, diversity and the microbial composition. However, the effect on the microbial biomass and diversity were not significant.

1. Introduction

For over half a century, bioleaching has been employed to economically extract metal from certain sulfide minerals [1]. In China, the first commercial heap bioleaching plant, with a capacity of 10,000 t.Cu/a, began operation at the Zijin Copper Mine at the end of 2005 [2].

The ZiJin Copper Mine is located in Fujian province in the southeast of China, which has richer rainfall and higher temperature than the rest of China. Due to the particular geographic location of the Zijin Copper Mine, heavy metals could be more easily to spread from the bioleaching heap to the nearby river with rainwater. In order to research and track of the safety of Zijin Copper Mine heap bioleaching plant, the periphery of the heap bioleaching plant was routine to test and risk assess from 2007.

To identify pollution problems, the anthropogenic contributions should be distinguished from the natural sources. Geochemical approaches, such as the geo-accumulation index (Igeo) [3, 4] is often used to distinguish anthropogenic contributions from the natural sources. Microorganisms are sensitive to heavy metals, and heavy metals have a significant influence on bacterial community structure, microbial biomass and microbial diversity [5], microbial ecology is an important indicator of heavy metal pollution. The application of molecular biology techniques, especially 16S ribosomal RNA high throughput sequencing technology, has progressed significantly in the field of microbial ecology [6].

In this paper, to research and track of the safety of Zijin Copper Mine heap bioleaching plant, the physical and chemical indexes of the water and sludge samples were also determined using ICP-OES, the extent and risks of heavy metal contamination were assessed by the Igeo method. Furthermore, the microbial community structure of samples from different environments were analyzed and compared using the 16S ribosomal RNA high throughput sequencing technology.

2. Materials and Methods

2.1. Sampling sites information

Ting River near Zijin Mining heap bioleaching plant (116°21′48.5″E; 25°12′13.2″N) located in Fujian province, belongs to subtropical maritime monsoon climate, the average temperature is 18.7~21.0 degrees, and the average rainfall is 1031~1369 mm.

2.2. Sample preparation and classification

19 sludge samples (S1-S19) and 11 water samples (W1-W11) were collected using sterile plastic bottles in October of 2015 year. All the samples were stored at 4℃ before determination. The spatial distribution of the sampling sites was shown in Figure 1.

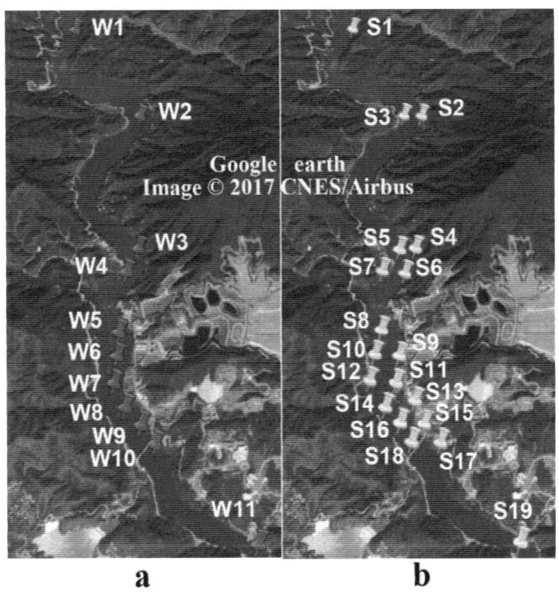

Figure 1. Sample collecting sites. Water samples collecting sites (a), Sludge samples collecting sites (b).

To convenient the analysis and research the effect of heap bioleaching plant on the nearby river, samples were divided into 3 groups according to their spatial distribution, and the details were shown in Table 1.

Table 1. Sample grouping information

Groups	Sludge samples	Water samples
Upstream	S1, S2, S3, S4, S5, S6, S7	W1, W2, W3, W4
Nearby	S8, S9, S10, S11, S12, S13, S14, S15, S16	W5, W6, W7, W8, W9
Downstream	S17, S18, S19	W10, W11

2.3. Physicochemical analysis of the samples
Sludge was filtered and dried thoroughly in the oven, grinded into powder. About 0.5g sludge powder was weighed, added 15mL nitric acid, 5mL hydrochloric acid and 2mL hydrochloric acid successively, heated until the liquid was evaporated completely and white smoke fumed, added 5mL deionized water and heated to boiling, added 5mL hydrochloric acid and heated to boiling, then cooled and added deionized water to 10mL[1,2]. The pre-treated sludge samples were filtered with super membrane filters (0.2 μm pore size, Sigma-Aldrich) and determined using ICP-OES[5]. The river water samples were determined using ICP-OES directly after filtration with super membrane filters (0.2 μm pore size, Sigma-Aldrich) [5].

2.4. Statistical analysis
The pH, Eh and heavy metals were analysed with separately one-way ANOVA (analyses of variance). All these statistical analysis were performed using SPSS 19.0 for Windows[7].

2.5. Geo-accumulation index analysis
The geo-accumulation index (*I*geo) was used to assess the extent of heavy metal pollution in sediments. This index was originally introduced by Müller as follows [3, 8]:

$$I_{geo} = \log_2(\frac{C_n}{1.5 \times B_n}) \quad (1)$$

Where C_n is the measured metal concentration of n, and B_n is the geochemical background concentration of metal n. In this study, the background value of soils in Fujian Province was used as the Bn (with Cr, Cu, Fe and Pb concentrations equalling 14 $mg \cdot kg^{-1}$, 22.8 $mg \cdot kg^{-1}$, 4.24 $mg \cdot kg^{-1}$, 41.3$mg \cdot kg^{-1}$, respectively)[9]. A value of 1.5 was used as the background matrix-correction factor to account for the lithogeny effect. The class and pollution status of samples were assessed according to table 2.

Table 2. Class and pollution status of samples by geo-accumulation indexes (Igeo)

Igeo	Class	Pollution Status	Risk
<0	1	Unpolluted	No risk
0-1	2	Unpolluted to moderate	Low risk
1-2	3	Moderate	Moderate risk
2-3	4	Moderate to heavy	Moderate risk to high risk
3-4	5	Heavy	High risk
4-5	6	Heavy to extreme	High risk to very high risk
>5	7	Extreme	Very high risk

2.6. DNA extraction, miseq sequencing and data processing
Whole DNA of samples was extracted using the E.Z.N.A. bacterial DNA kit (OMEGA, D3350-01) according to the manufacturer's instruction. 16S rRNA genes were sequenced with the 340F/805R primers [10]. Sequencing was conducted on an Illumina miSeq high throughput sequencing technology platform [11].
Paired-end reads of the original DNA fragments from high throughput sequencing were merged using FLASH[12] and assigned to each sample according to the unique barcodes. The 16S rRNA genes were processed using an open-source software QIIME [13]. Chimera Slayer tool was used for chimera detection[14], then CD-HIT package[15] and QIIME script "pick_de_novo_otus.py"[16] were used to pick operational taxonomic units (OTUs) by making OTU table, sequences with ≥ 97% similarity were assigned to the same OTUs[17]. Representative sequences for each OTU were picked and the RDP classifier was used to annotate taxonomic information for each representative sequence[18].

3. Results and Discussion

3.1. Physiochemical properties and spatial distribution of heavy metals in the water samples

Spatial distribution patterns and physiochemical properties of heavy metals in the river were shown in Table 3-4. The Eh and pH of water samples were in the range of 313-324 mV and 6.43-7.54, respectively. The Eh of the heap bioleaching plant nearby and downstream was obviously higher ($p<0.05$) than the upstream. The pH of the heap bioleaching plant nearby was slightly higher ($p>0.05$) than the upstream and downstream respectively. This was opposite to the results of 2009 [19] and consistent with the data of Longjiang river dosing sites [5]. This may be related to the use of a large amount of soda lime for remediation in 2012 pollution incidents.

Previous reports showed that S, Cu, Fe, Pb, As and Cr are the main elements of Zijin copper mine [20]. Thus, these six elements were determined to verify heap bioleaching plant pollution and influence on the nearby river. Soluble S concentration significantly decreased in the water samples followed the river direction, the spatial distribution difference of Pb in water was not significant and Cu could hardly be detected in water samples. However, the S, Pb and Cu content in downstream sludge notably increased compared with upstream sludge ($p<0.05$). The distribution of S, Pb and Cu illustrated there was no S, Pb and Cu discharged into the river now, some S, Pb and Cu migrated into the river from the heap bioleaching plant and was precipitated into the river sludge in the past time. The spatial distribution difference of As in water and the spatial distribution differences of Fe, Cr in sludge were not notable ($p>0.05$) along the river. Furthermore, As in sludge and Fe, Cr in water could hardly be detected. These results showed that heap bioleaching plant had no significant influence on the distribution of As, Fe and Cr.

The heavy metals concentration of water samples were compared with the current national standard of the People's Republic of China for surface water (GB3838-2002). Results showed that most of the water samples reached the third class standard of the People's Republic of China for surface water (GB3838-2002), and a small amount of water samples was above the fifth class standard of the People's Republic of China for surface water (GB3838-2002) because the high concentration of As [21].

Table 3. Physiochemical properties and distribution of heavy metals in water (mg/L)

Groups	pH	Eh	As	Pb	Soluble S
Upstream	6.92±0.44	315.5±3.11	0.1±0.02	0.03±0.01	30.28±5.94
Nearby	7.04±0.44	320±2.24	0.09±0.02	0.03±0.01	24.87±7.55
Downstream	6.46±0.04	321.5±3.54	0.09±0.00	0.03±0.00	6.67±2.07

Table 4. Spatial distribution of heavy metals in sludge (mg/kg)

Groups	Cr	Cu	Fe	Pb	S
Upstream	90.9±14.6	14.9±3.5	46685.7±4697.0	20.7±14.3	197.1±98.4
Nearby	93.3±23.1	87.3±102.9	45022.2±6635.7	34.2±22.3	390.2±576.3
Downstream	132±95.2	79.3±43.9	50633.3±22171.5	70.7±53.4	680±281.6

3.2. Geo-accumulation index analysis and pollution assessment of sludge samples

The geo-accumulation index was developed by Muller based on his study on heavy metals in fluvial stream sediments [8]. It has been widely used because it takes into account the effects of the natural elements and human activities on pollution [4]. The geo-accumulation indexes of Cr, Cu, Fe and Pb were calculated in this study (Fig. 2 and Table 5). Fe had the largest geo-accumulation index, Cr came the second, and the geo-accumulation indexes of Fe and Cr had no notable changes in all the sludge samples ($p>0.05$). These results indicated that the heap bioleaching plant had no obvious effect on the Fe and Cr accumulation in sludge, the high Igeo index of Fe and Cr result from the natural sources rather than the anthropogenic contributions from heap bioleaching plant. The geo-accumulation index of Cu differed significantly ($p<0.05$), which in the heap bioleaching plant nearby and downstream samples increased notably than in upstream samples. Most of the geo-accumulation indexes of Pb were less than 0, only a few of samples from the heap bioleaching plant nearby and downstream were more than 0 and less than 1. These results implied that the anthropogenic contributions from heap bioleaching plant increased the Cu and Pb pollution of the sludge samples. However, the effect of heap bioleaching plant on Pb was subtle.

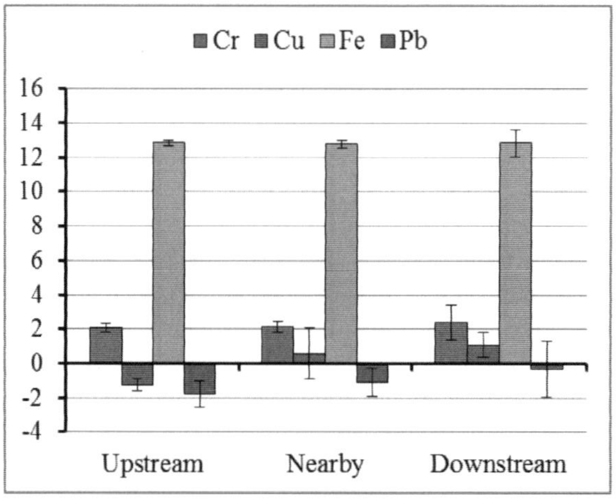

Figure 2. Geo-accumulation indexes (Igeo) of heavy metals in sludge samples

Table 5. Pollution status statistical table of various heavy metals

Class	Pollution Status	Number of samples				Percentage			
		Cr	Cu	Fe	Pb	Cr	Cu	Fe	Pb
1	Unpolluted	0	11	0	15	0.0	57.9	0.0	78.9
2	Unpolluted to moderate	0	5	0	4	0.0	26.3	0.0	21.1
3	Moderate	6	1	0	0	31.6	5.3	0.0	0.0
4	Moderate to heavy	12	1	0	0	63.2	5.3	0.0	0.0
5	Heavy	1	1	0	0	5.3	5.3	0.0	0.0
6	Heavy to extreme	0	0	0	0	0.0	0.0	0.0	0.0
7	Extreme	0	0	19	0	0.0	0.0	100.0	0.0

3.3. Microbial diversity and composition of the samples

Microbial biomass (richness) and diversity (Shannon index) results of the samples were shown in Fig.3. In the sludge samples, the microbial biomass and diversity in the downstream samples lowered than the other two groups. In the water samples, the microbial biomass and diversity decreased followed the river. However, the difference was not notable ($p>0.05$). These results indicated that the heap bioleaching plant had some effect on the microbial biomass and diversity. However, the effect of heap bioleaching plant on microbial biomass and diversity was not significant.

Microbial community structures results of different samples were shown in Fig. 4. There were some *Leptospirillum* existed in some of the upstream samples, this is consistent with the previous reports [19]. These results indicated the *Leptospirillum* is the inherent microbe of this river. More *Leptospirillum* and *Thiobacillus* existed in the the heap bioleaching plant nearby and downstream samples than in the upstream samples implied the heap bioleaching plant influenced on the microbial community structures. There are notable microbial species differences between the water and sludge samples. *Sulfuricurvum* could hardly be detected in the sludge samples, *Sulfuricurvum* was the dominant microbe in all the river water samples, and the difference in all the water samples was not notable. These results indicated *Sulfuricurvum* mainly existed in the water, and *Sulfuricurvum* was also the inherent microbe of the river water.

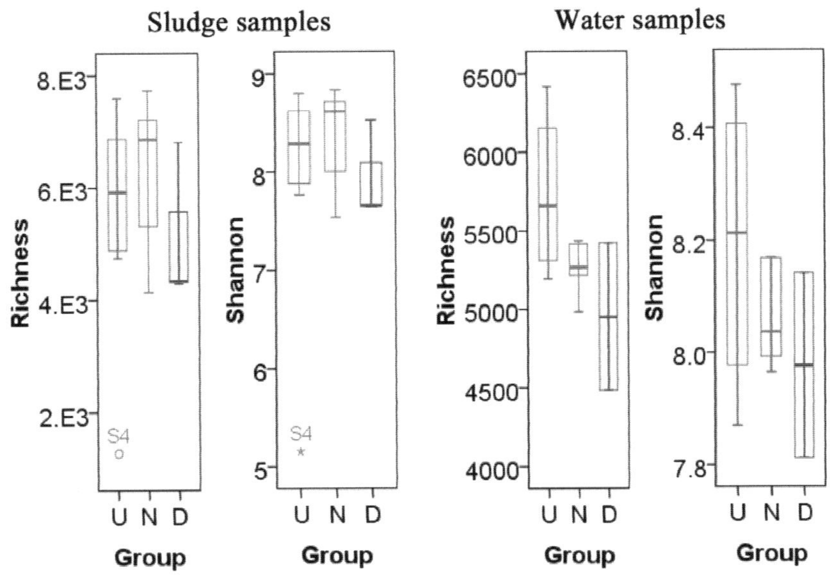

Figure 3. Microbial biomass and diversity of different samples (Upstream-U; Nearby-N; Downstream-D)

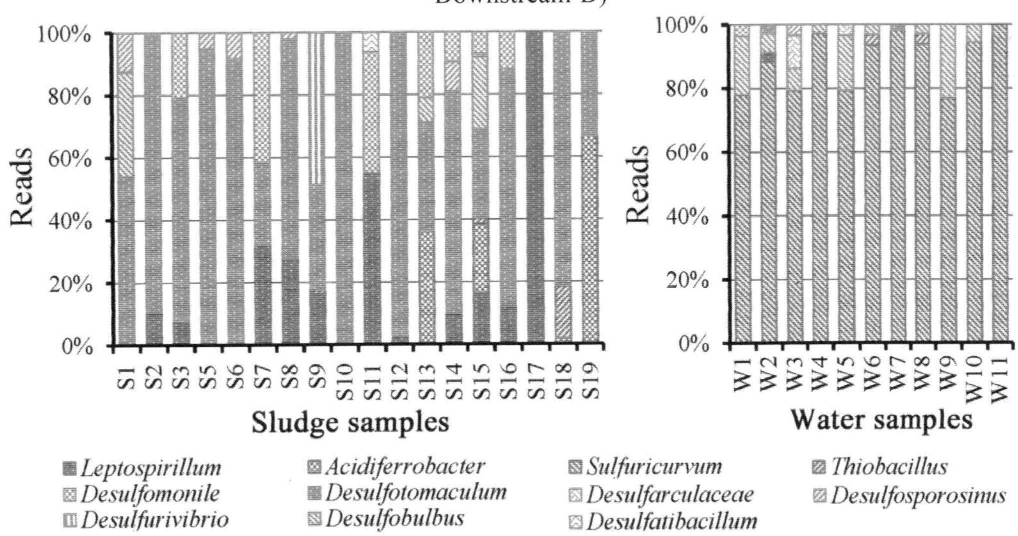

Figure 4. microbial community structures of different samples

4. Conclusions

The heap bioleaching plant had significant influence on the distribution of S, Pb and Cu and no significant influence on the distribution of As, Fe and Cr. Most of the water samples reached the third class standard of the People's Republic of China for surface water and individual water samples were above the fifth class standard of the People's Republic of China for surface water (GB3838-2002) because of As. The heap bioleaching plant had some effect on the microbial biomass, diversity and the microbial composition, however, the effect on the microbial biomass and diversity were not significant.

5. References

[1] Brierley C and Brierley J 2013 Progress in bioleaching: part B: applications of microbial processes by the minerals industries *Applied Microbiology and Biotechnology* 97 7543

[2] Ruan R, Wen J and Chen J 2006 Bacterial Heap-Leaching: Practice in Zijinshan Copper Mine *Hydrometallurgy* 83 77
[3] Zhang L, Liao Q, Shao S, Zhang N, Shen Q and Liu C 2015 Heavy Metal Pollution, Fractionation, and Potential Ecological Risks in Sediments from Lake Chaohu (Eastern China) and the Surrounding Rivers *Int J Environ Res Public Health* 12 14115
[4] Song J, Yang X, Zhang J, Long Y, Zhang Y and Zhang T 2015 Assessing the Variability of Heavy Metal Concentrations in Liquid-Solid Two-Phase and Related Environmental Risks in the Weihe River of Shaanxi Province, China *Int J Environ Res Public Health* 12 8243
[5] Zhang M, Huang F, Wang G, Liu X, Wen J, Zhang X, Huang Y and Xia Y 2017 Geographic distribution of cadmium and its interaction with the microbial community in the Longjiang River: risk evaluation after a shocking pollution accident *Sci Rep* 7 227
[6] Amann R I, Ludwig W and Schleifer K H 1995 Phylogenetic identification and in situ detection of individual microbial cells without cultivation *Microbiol. Rev.* 59 143
[7] Li J, Zheng Y, Yan J, Li H, Wang X, He J and Ding G 2013 Effects of different regeneration scenarios and fertilizer treatments on soil microbial ecology in reclaimed opencast mining areas on the Loess Plateau, China *PLoS One* 8 e63275
[8] Muller G 1969 Index of geoaccumulation in sediments of the Rhine River *Geojournal* 2 108
[9] Wei F, Chen J and Wu Y. Background Element Values in Soils of China. 1990, Beijing(china): China Environmental Science Press. 329.
[10] Sinclair L, Osman O A, Bertilsson S and Eiler A 2015 Microbial community composition and diversity via 16S rRNA gene amplicons: evaluating the illumina platform *PLoS One* 10 e0116955
[11] Unno T 2015 Bioinformatic Suggestions on MiSeq-based Microbial Community Analysis *J Microbiol Biotechnol*
[12] Magoc T and Salzberg S L 2011 FLASH: fast length adjustment of short reads to improve genome assemblies *Bioinformatics* 27 2957
[13] Kuczynski J, Stombaugh J, Walters W A, Gonzalez A, Caporaso J G and Knight R 2012 Using QIIME to analyze 16S rRNA gene sequences from microbial communities *Curr Protoc Microbiol* Chapter 1 Unit 1E 5
[14] Haas B J, Gevers D, Earl A M, Feldgarden M, Ward D V, Giannoukos G, Ciulla D, Tabbaa D, Highlander S K, Sodergren E, Methe B, DeSantis T Z, Human Microbiome C, Petrosino J F, Knight R and Birren B W 2011 Chimeric 16S rRNA sequence formation and detection in Sanger and 454-pyrosequenced PCR amplicons *Genome Res* 21 494
[15] Li W, Fu L, Niu B, Wu S and Wooley J 2012 Ultrafast clustering algorithms for metagenomic sequence analysis *Brief Bioinform* 13 656
[16] Caporaso J G, Kuczynski J, Stombaugh J, Bittinger K, Bushman F D, Costello E K, Fierer N, Pena A G, Goodrich J K, Gordon J I, Huttley G A, Kelley S T, Knights D, Koenig J E, Ley R E, Lozupone C A, McDonald D, Muegge B D, Pirrung M, Reeder J, Sevinsky J R, Turnbaugh P J, Walters W A, Widmann J, Yatsunenko T, Zaneveld J and Knight R 2010 QIIME allows analysis of high-throughput community sequencing data *Nat Methods* 7 335
[17] Li W and Godzik A 2006 Cd-hit: a fast program for clustering and comparing large sets of protein or nucleotide sequences *Bioinformatics* 22 1658
[18] Wang Q, Garrity G M, Tiedje J M and Cole J R 2007 Naïve bayesian classifier for rapid assignment of rrna sequences into the new bacterial taxonomy *Applied and environmental microbiology* 73 5261
[19] Liu X, Chen B, Chen J, Zou L, Zhang M, Wen J and Liu W 2016 Biogeographical distribution of acidophiles and their effects around the Zijinshan heap bioleaching plant *Chemistry and Ecology* 32 419
[20] Liu X, Chen B, Chen J, Mingjiang Z, Jiankang W, Dianzuo W and Renman R 2016 Spatial variation of microbial community structure in the Zijinshan commercial copper heap bioleaching plant *Minerals Engineering* 94 76
[21] Administration C S E P. (2002) standard of the People's Republic of China for surface water Beijing.

Acknowledgments
This work was supported by the National Natural Science Foundation of China under grant numbers U1402234, 41573074; the nation high-level youth talents special support plan; the Guangxi scientific research and technology development plan under grants GuikeAB16380287 and GuikeAB17129025; the public welfare fund of the Ministry of Environmental Protection of People's Republic of China under grant number 201509049; Program of International S & T Cooperation 2016YFE0130700; and the fund of General Research Institute for Nonferrous metals under grant numbers 53321 and 53348.

IOP Conf. Series: Earth and Environmental Science **182** (2018) 012006 doi:10.1088/1755-1315/182/1/012006

Fraction Transformation of Cr in *Leersia hexandra* Swartz Constructed Wetland

Jiansheng Wang [1,2], Xingfeng Zhang [1,2], Fengjiao Gao [1,2] and Masafumi Goto[3]

1 Guangxi Key Laboratory of Guangxi Environmental Pollution Control Theory and Technology, Guilin University of Technology Guilin 541004， China
2 Collaborative Innovation Center for Water Pollution Control and Water Safety in Karst Area, Guilin University of Technology, Guilin 541004, China
3 Malaysia-Japan International Institute of Technology (MJIIT), Universiti Teknologi Malaysia Kuala Lumpur, Kuala Lumpur 54100, Malaysia
E-mail: wjs_0506@163.com

Abstract. In this paper, the fraction distribution of Cr contaminated wastewater were treated by *Leersia hexandra* Swartz constructed wetlands. The results showed that with the Cr-contaminated wastewater enter the wetland, the wastewaters transformed to different forms respectively, and residuals fraction mainly were existed in the matrix and in plant bodies. The residuals present in all zones turned up at first and then decreased; the reducible fraction showed a gradual decreased in the wetland; and the weak acid extractable did not change much. The fraction of Cr in *L. hexandra* roots increased first and then decreased in different zone, of which the most obvious changes were in weak acid extractable fraction. The fraction of weak acid extractable and residue in the stems of *L. hexandra* increased first and then decreased in different zone, while the ethanol fraction decreased gradually. Residue fraction and ethanol fraction in different zone gradually decreased in the *L. hexandra* leaves, while the weak acid extractable showed the first increase and then decrease.

1. Introduction

After heavy metals got into the soil through a series of physical and chemical reactions, the formation of different chemical forms, and showed different activities. The biological toxicity of heavy metals was not only related to their total amount, but also to a greater extent determined by their morphological distribution. Different forms of heavy metal had different environmental effects, which directly affected the toxicity and migration of heavy metals and the circulation in nature. The form and transformation of heavy metals in soils was of great significance for studying the environmental effects of heavy metals and the remediation of soils contaminated by heavy metals[1-2].

Hexavalent chromium [Cr (VI)] was considered a hazardous water pollutant. Thus, the removal of Cr (VI) was necessary for the protection of the environment [3]. The two common oxidation fractions of Cr were present in the environment; i.e. Cr (III) and Cr (VI).Cr (III) was considered to be a trace element essential for the proper functioning of living organisms. It was reported to be responsible for the control of glucose and lipid metabolism in mammals [4-5]. Cr (III) was difficult to dissolve in water and considered stable in natural environment. Cr (IV) was highly water soluble and known to be highly toxic including carcinogenicity [6].

Leersia hexandra Swartz (*L. hexandra*) was grass, which grown in temperate regions, propagates very rapidly with high growth density and biomass per unit area. The plant was also known to be capable of accumulating a variety of heavy metals[7].Constructed wetlands (CWs) were engineered biogeochemical systems that mimic the natural wetland in remediation process, where a complex system of plants, microorganisms, and soil substrate act together to treat contaminated water. Constructed wetlands had the advantages of high efficiency, low investment, low running cost, was widely used in the treatment life, a variety of types such as industrial wastewater, was developing rapidly in recent years, ecological wastewater treatment technology [8].

2. Materials and Methods

2.1. The Experimental Materials

2.1.1. Plant. The *L. hexandra* used in this study were obtained from the paddy field of Yanshan Campus of Guilin University of Technology, Guilin, China.

2.1.2. Experimental Constructed Wetland. A constructed wetland with the dimensions of 1.3 m (L) ×0.7 m (W) ×0.5 m (H) was used in the study.
The length of the wetland inflow area was 0.1 m, the length of the plant area is 0.4 m, each for the following three compartments (plant areas). Each compartment was filled with gravels in 0.1 m thickness on the bottom, then topped with 0.3 m Thick layer of red soil. Cut view of the constructed wetland was depicted in Figure 1.
Gravels, which were obtained from a construction team in Guangxi, have the particle size of 30-60 mm, and the red soil, which was collected from the campus of Guilin University of Technology, has particle size of 0.9 - 3.9 mm.
Overall layout of the CW is shown in Figure 1. As shown in Figure 1, the constructed wetlands consists of water inlet zone, three plant areas number 1, 2 and 3 and an outlet; Zone 1 was the wetland primary stage, zone 2 was the wetland middle stage and zone 3 was the wetland last stage. What flow in the CW was shown by arrow. A, B, C and D in the figure were the sampling ports for water samples designated as No.1 effluent, No. 2 effluent, No.3 effluent and the final effluent, respectively. The influent to the CW was an artificial Cr-polluted wastewater with Cr (VI) concentration of 20 mg/L.$K_2Cr_2O_7$, was used in preparation of the artificial wastewater.

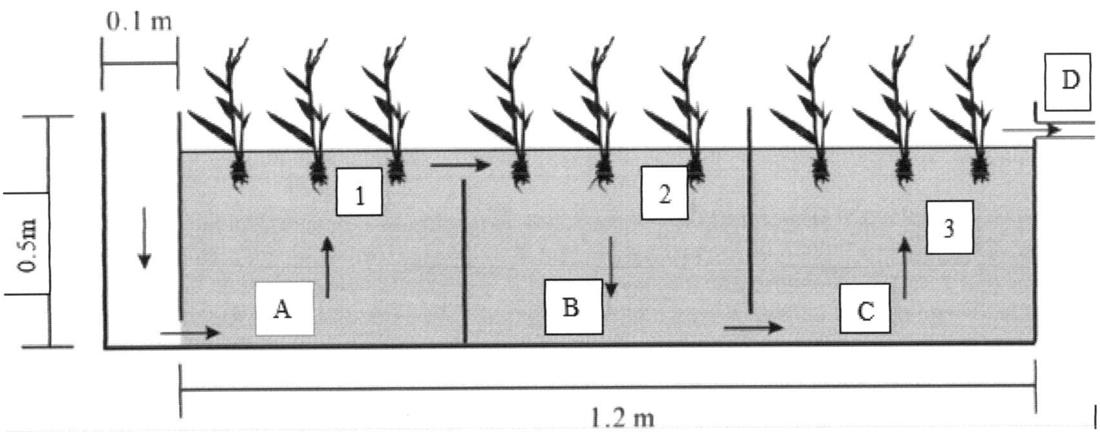

Figure 1. Schematic of wetland setup used in this study. The arrow represents direction of water flow.

2.2. Method

2.2.1. Collection and treatment of plant samples. After 180 days of normal operation of the constructed wetland, the *L. hexandra* were harvested, and the roots, stems and leaves of the plants were collected. The collected plants were repeatedly rinsed with tap water, and washed three times with ultrapure water, then sonicted for 15 min in an ultrasonic cleaner, finally washed three times with deionized water in order to remove trace amount of metal attached on the plant bodies. The washed plant samples were divided into three parts: root, stem and leaf. Each part was deactived at 105 °C for 30 minutes and then dried to constant weight at 70 °C.

2.2.2 Collection and treatment of soil sample

Soil samples were collected in the CW for 180 days, and air-dried in a cool place for more than fifteen days until constant weight, and sieved by 100 mesh sieves.

2.3. Analysis of Cr in the Wetland Matrix

Cr extraction of matrix was constructed by the sequential BCR extraction method, and the acid extractable chromium content was determined in each sample as follows [2].

(1) 1.00 g of prepared soil sample in a cetrifuge tube (100 mL) was added with 40 mL of 0.11 M CH_3COOH, then the tubers were shaken by a reciprocal shaker at 250 rpm and at 25 °C for 16 h. The centrifuge tubes were centrifuged at 4,000 rpm for 15 min, and each supernatant was transferred to a 50 mL volumetric flack. Ultrapure water was added to bring the volume to 50.0 mL. The concentration of Cr in the diluted sample was determined by atomic absorption spectrophotometer. 20 mL of distilled water was added to the residual soil sample, shaken for 15 minutes, and centrifuged at 4,000 rpm for 15 minutes. Then, water phase was discarded,and the remaining solid sample was kept for the next analysis.

(2) Determination of chromium content in reducible (oxidation bound) state. The remaining soil samples from above procedure (1) were added with 40 mL mixture of 0.5 M hydroxylamine hydrochloride and 0.05 M HNO_3, and shaken on a reciprocal shaker at 250 rpm and at 25 °C for 16 h, and then centrifuged at 4,000 rpm for 15 min. Finally, the supernatant in the centrifuge tube was transferred to a 50 mL volumetric flask and diluted with ultrapure water to 50.0 mL. Cr concentration was determined by atomic absorption spectrometry. Residual solid sample was kept for the next analysis.

(3) Determination of chromium content in oxidation (organic combination) state. 10 mL of H_2O_2 was added to the remaining soil sample, pH was adjsuted to 2 ~ 3 with HNO_3, then the tube was left at room temperature for 1 h (with occasional stirring by a glass rod), then kept on 85 °C water bath for 1 h with a cap. Heating continued without cap until the volume was reduced to no more than 3 mL. then 10 mL of H2O2 was added to the tube for additional heating on the water bath for 1 h with cap, then without cap until the volume was reduced to 1 mL. After cooling, the mixture was added with 1.0 M ammonium acetate solution (pH = 2.0) to 50mL, shaken continuously for 16 h, and finally centrifuged to obtain binding form Cr test solution.

(4) Determination of residual state content. Used triacid nitrification to analysis and test the total amount for the after determination solution, Subtract respectively the acid extractable content, oxidation state content and organic binding state content is the residual content.

2.4. Analysis of Cr in the Wetland Plants

The chemical forms of Cr in the *L. hexandra* sample specimens grown in the constructed wetland were studied by two-step continuous extraction method.

(1) Ethanol extractable Cr fraction. 1.0 g dry plant sample was put into 100 mL centrifuge tube and added with 10 mL 80% ethanol. The prepared tube were shanken continuously for 16 h at 25 °C and at 160 rpm, then centrifuged at 4,000 rpm for 10 min. The supernatant was transferred to a 150 mL Erlenmeyer flask, and then remaining residue was added with 10 mL of ethanol extractant, shanken for 2h under the same conditions as above, then centrifuged for 10 min to obtain supernatant which was added to the Erlenmeyer flask. This procedure was repated twice or three times and all the supernatants obtained were transferred to the 150 mL Erlenmeyer flask , and heated and concentrated on a hot plate at 140 °C for 2 h. A small funnel on the Erlenmeyer flask was used for a reflux effects; 2.0 mL of concentrated HNo3 was added during the concentration. When the solution was concentrated to near dryness, the Erlenmeyer flask was removed from the hot plate and cooled to room temperature, then added with deionized water to a 25.0 ml and transferred to a cuvette for analysis.

(2) Hydrochloric acid extractable Cr fraction. The residue from procedure (1) was added with 10.0 mL of 0.6 M HCL extractant followed by the same extraction procedure as described in (1).

(3) Residue state Cr fraction. The residue of procedure (2) was digested in HNO3-H2O2 system.

The contents of heavy metals in each prepared sample were determined by ICP and atomic fluorescence spectrometry.

3. Results and Discussion

3.1. Cr matrix form Distribution in Constructed Wetlands

The distribution of Cr in various forms in the CW was shown in Figure 2. With the operation of the CW, chromium-containing wastewater entered into the wetland through the gravel area in Zone 1and flows through the CW subject to a series of functions of the wetland matrix, plants and microorganisms (Figure 1). As a result of these functions, a part of chromium in the wastewater was removed by precipitation, adsorption and absorption.

As can be seen in figure 2, the fraction of residual Cr in the CW was the highest among the four fractions investigated, which implies that the aqueous chromium in the wastewater entering the CW could be adsorbed, complexed and/or transformed into the forms other than aqueous by the mineral particles in the soil. The acid extractable fraction, reducible fraction and oxidizable fraction were all less than 20% or less of the fraction. As shown in the Figure 2, the reducible form of chromium in Zone 2 had increased. Other forms of Cr in Zones 1, 2 and 3 had basically decreased as the wastewater flowed throught the CW. The reduction of the hydrochloric acid extractable fraction of Cr is considered to be related to the change between oxidizable and reducible and environment. With the operation of chromium-bearing heavy metal wastewater in CWs, the heavy metal-containing wastewater entering the CWs changed in the matrix of constructed wetland. Except for the reducible fraction of chromium heavy metal of zone 2 had an increase than zone 1, other forms of Cr content of heavy metals were Zone 1> Zone 2> Zone 3. The reduction of the fraction of heavy metal chromium-weak acid extractable bound fraction were related to the changes of oxidizable and reducible morphological contents and the environment.

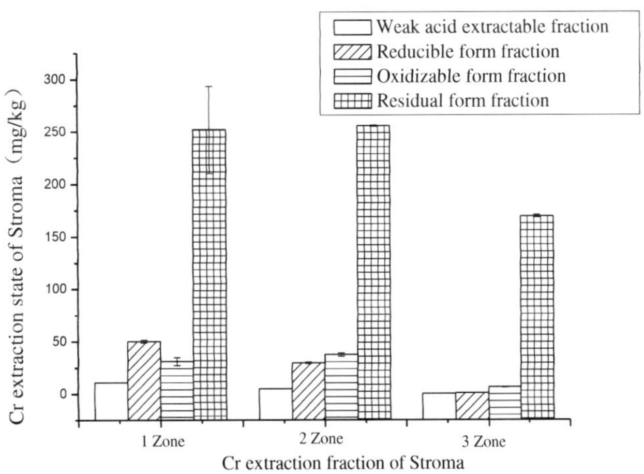

Figure 2. Distribution of Cr in the Constructed Wetland Material

3.2. The Distribution of Cr in Constructed Wetlands Plants

The distribution of Cr with in the roots of the plant grown in the CWs is shown in Figure 3 a. In the wetland, the distribution of Cr in the roots of *L. hexandra* was in the order of weak acid extractable fraction > residual fraction> ethanol fraction. Fractions of Cr in the roots ethanol, hydrochloric acid extractable, and residual forms in the CW plants increased from Zones 1 to 2 and then decreased from Zones 2 to 3, especially, changes in ethanol fractions are obvious(Figure 1 a).

The distribution of Cr in the stems of *L. hexandra* grown in the CW observed is shown in Figure 3 b. It can be seen from the figure, fraction of Cr in the stems of *L. hexandra* were the highest in Zone 2 in the three forms studied. It was also observed that Cr fraction in the stems were in the order of residual > ethanol extractable > hydrochloric acid extractable fractions in each zone. In all the forms of Cr studied, the fraction of Cr in the stems of *L. hexandra* was highest in the residual form in Zone 2 (Figure 3 b). In the CW study with the influent of chromium contained wastewater, the Cr distribution data that Zone 2 had the higher Cr fractions than Zone 1 and 3 in the sequence. The content of residual and hydrochloric acid extractable fractions increased from Zone 1 to 2,and the decreased from Zone 2 to 3 like in Cr in the stems. While the ethanol extractable fractions decreased from Zone 1, 2 and 3 in this sequence.

The distribution of Cr in the leaves of *L. hexandra* is shown in Figure 3 c. From the figure, it can be seen that residual Cr in the leaves of *L. hexandra* grown in a zone of CW was more than that hydrochloric acid extractable and ethanol extractable fractions in the same zone. The residual fraction was the highest in Zone 1 and decreased as the wastewater flowed in Zone 2 and Zone 3. Residual form was found present in Zone 1 accounting for 88.14% of the total amount of residual form of Cr in the leaves. The residual fraction of Cr in the leaves decreased from in Zone 1 to in the Zone 2 and Zone 3. The content of total Cr form to district 57.80%, hydrochloric acid extractable form increased in Zone 2 compared with that in Zone 1indicating Cr absorbed by *L. hexandra* leaves in Zone 2 was transformed-into hydrochloric acid extractable state, while ethanol extractable content and residual form of Cr in Zone 1 decreased in Zones 2 and 3. Only hydrochloric acid extractable fraction increased in Zone 2 from Zone 1, but decreased in Zone 3.

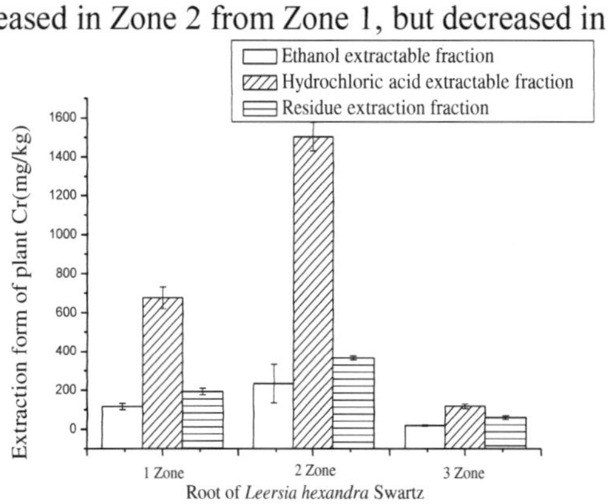

a Root

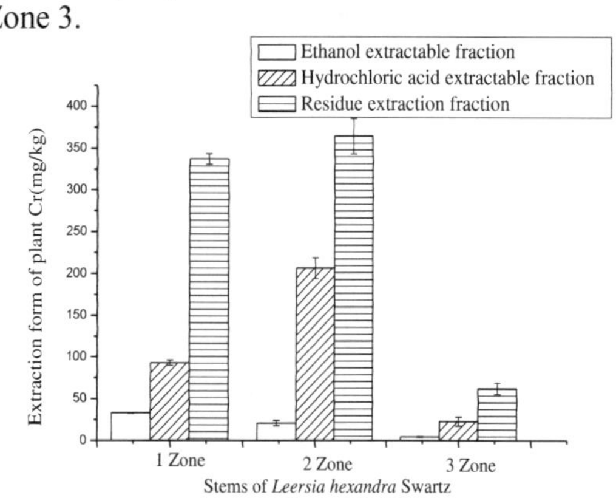

b Stems

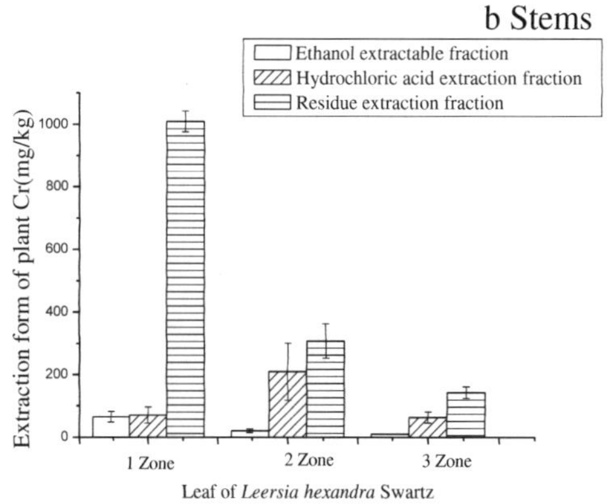

c Leaves

Figure 3. Distribution of Cr in the *L. hexandra* grown in the CW

It was observed in the study that Cr in the CW was mainly in residual form, followed by the reducible and oxidizable forms. These results are in agreement with the data reported by Zhang Xinyan[9] on Cr behaviors in CW. Cr contaminated wastewater into the soil, Cr reducible content with the wetland operation, its concentration decreased. The research by You Shaohong[10] showed that in the CW to maintain the stability of the system had played an important role. At the same time, it was also observed that pH, organic matter and microbial population in the CW system all affect on the accumulation of Cr in the CW matrix. It has been investigated and reported by Zhang[9] that addition of humic acid into the CW matrix enhances significantly transfer of exchangeable and carbonate forms to organic fraction.

4. Conclusion

In the construction wetland, the residual form of Cr mainly in the CW matrix, while the presence weak acid extractable forms of Cr changes little in the CW. In the parts of *L. hexandra* grown in the CW over the period of 180 days, the hydrochloric acid extractable form of Cr in the roots is in the hightest amount, while Cr in the residual state is mainly in the stems and leaves. It is concluded that *L. hexandra* in the CW can increase Cr accumulation in the matrix and plant resulting in an enhanced removal of aqueous Cr in the influent.

5. References

[1] Leea C G, Leeb S, Parka J Ann, Parka C, Leea S J, Kimc S B, Byungryul, Yunb S T, Leea S H and Choia J W 2016 Removal of copper, nickel and chromium mixtures from metal plating wastewater by adsorption with modified carbon foam *J.Chemosphere*.166 203-211.

[2] Han C M,Wang L S, Gong Z Q and Xu H X 2005Chemical forms of soil heavy metals and their environmental significance *J.Chinese Journal of Ecology*. 24(12) 1499-1502.

[3] Liu Jie, Zhang Xuehong, You Shaohong, Wu Q X, Ceng S M and Zhou K N 2014 Cr (VI) removal and detoxification in constructed wetlands planted with *Leersia hexandra* Swartz *J.Ecological Engineering*.71 36-40.

[4] Carlos B D, Violeta L L and Bilyeu B 2012 A review of chemical, electrochemical and biological methods for aqueous Cr (VI) reduction *J.Journal of Hazardous Materials*. 223-224 1-12.

[5] Dhal B, Thatoi H N, Das N N and Pandey B D 2013 Chemical and microbial remediation of hexavalent chromium from contaminated soil and mining/metallurgical solid waste: A review *J.Journal of Hazardous Materials*.250-251 272-291.

[6] Kotaś J and Stasicka Z 2000 Chromium occurrence in the environment and methods of its speciation *J.Environmental Pollution*. 107 263-283.

[7] Zang X H, Luo Y Q, Huang H T, Liu J, Zhu Y N and Zheng Q F 2006 *Leersia hexandra* Swartz: a newly discovered hygrophyte with chromium hyper-accumulator properties *J.China Acta ecologica sinica*.26 950-953.

[8] Arivoli A, Mohanraj R and Seenivasan R 2015 Application of vertical flow constructed wetland in treatment of heavy metals from pulp and paper industry wastewater *J.Springer Berlin Heidelberg*. 22 13336-43

[9] Zhang Xinyan 2014 Study on Transfer and Accumulation of Chromium (Cr) in Constructed Wetland Systems wih Coix aquatica Roxb *Guangxi University*

[10] You Shaohong 2017 Decontamination performance and mechanism of Cr (VI) polluted water using *Leersia hexandra* Swartz constructed wetland planted with bagasse biochar soil matrix. Nanning: *Guangxi University*

Acknowledgments

The Guangxi Science Technology Major Project (GuikeAA17204047), the Special Funding for Guangxi “Specially-invited Expert” and the Guangxi Scientific Experiment Center of Mining, Metallurgy and Environment（KH2012ZD004）

IOP Conf. Series: Earth and Environmental Science **182** (2018) 012007 doi:10.1088/1755-1315/182/1/012007

Application and Analysis of Bayesian Method and Grey Relational Analysis in Marine Water Quality Evaluation

WenchaoZhang[1], HuiyingGao[1], HaiSun[1]

1 Ocean University of China, Qingdao, China
E-mail: 18306485875@163.com

Abstract. There are many factors that influence the quality of marine water. In order to make the evaluation process more efficient and accurate, based on the normal distribution principle and Bias formula, this article establishes the seawater quality evaluation model by Bayesian method. Taking the evaluation of water quality of a sea area in Qingdao as an example, the measured data of eight water quality monitoring points are selected. The results of the evaluation are compared with the grey relational analysis. It shows that the results of the two methods are the same, which are the I type of water quality. So the Bayesian method based on normal distribution is applicable to the evaluation of marine water quality and the Bayesian method has the characteristic of more integrated, suitable for both large and small samples, simple calculation and easily to be used widely.

1. Introduction

With the development of the marine economy, the environmental conditions of marine waters are becoming more and more severe. People's awareness of environmental protection has been gradually improved. The evaluation of marine water quality is of great significance to the improvement of the marine environment. Many researchers have studied the methods of marine water quality evaluation. There are many uncertainties. Considering the fuzziness and grey nature of evaluation, some evaluation methods such as fuzzy mathematics and grey relational analysis are put forward by some researchers, which have achieved certain results in the application of marine water quality evaluation. However, these methods generally have the characteristics of large calculation and complex calculation, and can not meet the requirements of high aging calculation for small samples [1].

In water quality evaluation, we should avoid the influence of subjective factors as far as possible. Bayesian method is widely used to solve the problem of uncertainty, this method can avoid the influence of excessive subjective factors.

In this article, the normal distribution principle and Bayesian method are combined to establish the marine water quality evaluation model, and compared with the grey relational analysis method. The two evaluation results are compared, in order to provide a more effective and convenient new method for marine environmental assessment.

2. The Principle of Bayesian Method and Grey Relational Analysis Method

2.1. *Bayesian Method*

Bayesian method is based on existing data, using probability theory and statistics method to predict and analyse an event, that is, to guess the probability of this event. Bayesian method originated in a paper in 1763, mainly based on Bayes' theorem, which can be used to solve the problem of statistics and the probability of events[2] [3]. Bayesian method is combined with probability theory and mathematical statistics method.

Bayesian method can be used in water quality evaluation. In order to avoid the sampling error, principle of normal distribution is used to calculate the likelihood probability. This method has the advantage of more integrated, suitable for both large and small samples and its requirements for data is low, with less effort[4].

2.2. Grey Relational Analysis Method

Grey relational analysis is a quantitative analysis of uncertain relations between different factors. This method analysis comparative sequence and reference sequences to judge whether associated closely. Grey relational analysis is often used in the environmental evaluation, the monitoring of sample data as a reference sequence, different standard of environmental indicators as a comparative sequence. The method compares correlation degree, a more similar trend means a closer relationship. This method can reduce the loss caused by the information asymmetry [5].

3. The Steps of Bayesian Method and Grey Relational Analysis method

3.1. The Steps of Bayesian Method

3.1.1. Improved the Bayesian formula

To solve the problem of water quality evaluation, the Bayesian formula can be improved.

$$P(y_{ij}|x_i)=\frac{P(y_{ij})P(x_i|y_{ij})}{\sum_{i=1}^{m}P(y_{ij})P(x_i|y_{ij})} \quad (1)$$

Where y_{ij} is sea water quality standards, i is the evaluation index, i=1,2……m; and j is seawater standard type, j= I, II, III, IV; x_k is monitoring indexes, the number of monitoring is n.

$P(y_{ij})$ is priori probability, it represents the possibility of the water quality level speculated by the experience.

$P(x_i|y_{ij})$ is likelihood function, it represents the water quality value of the monitoring point. According to the statistical theory, when water quality belongs to the grade j, there is a certain sampling error between the sample value and the standard value, and the distribution can be expressed in the normal distribution.

$P(y_{ij}|x_i)$ is posteriori probability, it indicates that water quality is the possibility of grade j under the condition of $P(y_{ij})$. The evaluation of water quality is usually based on the posterior probability. Based on this consideration, this paper uses the principle of normal distribution of sampling error to estimate the prior probability [6].

3.1.2. A Priori Probability Calculation

It is assumed that the water quality of each monitoring point is the same as the probability of a certain index

$$P(y_{i\,\mathrm{I}}) = P(y_{i\,\mathrm{II}}) = \cdots \ldots = P(y_{i\mathrm{IV}})=\frac{1}{4} \quad (2)$$

3.1.3. Calculation

The calculated likelihood function $P(x_i|y_{ij})$ is the probability that a single index is calculated to belong to a certain level. Estimation of likelihood function by the principle of normal distribution. The mean value of water quality standard value of i index is taken as an α_i of the normal distribution of the index, and the standard difference of water quality of i is σ_i, and C_{vi} is the coefficient of variation of i index. Estimate C_{vi} from $C_{vi}=\frac{\sigma_i}{\alpha_i}$.

Calculation of standard deviation α_{ij}, $\alpha_{ij}=C_{vi}\,y_{ij}$ for class j index.

The calculated value of the σ_{ij} into the $t_i=\frac{xi-yij}{\sigma ij}$ is standardized, and the standardized normal distribution is used to calculate the $P(x_i|y_{ij})$.

$$P(x_i|y_{ij})=2(1-\varphi(|t_i|)) \tag{3}$$

$$\varphi(|t_i|)=\int_{-\infty}^{t_i}\frac{1}{\sqrt{2\pi}}e^{\frac{u}{2}}du \tag{4}$$

According to the calculation results of $P(x_i|y_{ij})$, the posterior probability $P(y_{ij}|x_i)$ of a single index is calculated by $P(y_{ij}|x_i)=\frac{P(y_{ij})P(x_i|y_{ij})}{\sum_{i=1}^{m}P(y_{ij})P(x_i|y_{ij})}$.

3.1.4. Multi index comprehensive water quality test probability

$$Pj=\sum_{i=1}^{m}w_i\,P(y_{ij}|x_i) \tag{5}$$

W_i is the weight of different water quality index i, and can be determined according to the influence of various water quality indexes on the water quality.
In the comprehensive evaluation of water quality, the average weight is taken, that is, $W_{i1}=W_{i2}=...=W_{i8}=\frac{1}{9}$.

3.1.5. Evaluate the water quality

The results of water quality evaluation are determined according to the maximum probability principle. That is, the greater the probability, the greater the possibility of a certain category of water.

3.2. The Steps of Grey Relational Analysis method

3.2.1. Transform the original data, build a reference sequence and a comparison sequence

The monitoring data constitute the reference sequence x ,

$$x_i^0=\{x_i^0(1),x_i^0(2),\cdots\cdots,x_i^0(n)\},(i=1,2,\cdots\cdots,m).$$

Using the average method, eliminate the differences in order of magnitude

$$x_i=\{x_i(1),x_i(2),\cdots\cdots,x_i(n)\}=\{\frac{x_i^0(1)}{\overline{x_i^0}},\frac{x_i^0(2)}{\overline{x_i^0}},\dots\dots,\frac{x_i^0}{\overline{x_i^0}}\},(i=1,2,\cdots\cdots,m) \tag{6}$$

$$\overline{x_i^0}=\frac{1}{n}\sum_{k=1}^{n}x_i^0(k) \tag{7}$$

3.2.2. Calculate the grey relational coefficient

Nondimensionalization the original data sequence to normalize the data before processing. (n is the number of evaluation factors, m representative sample quantity)
x_j (k) represents the monitoring result of k at the monitoring stations of i

$$x_i=\{x_i(1),x_i(2),\cdots\cdots,x_i(n)\}_j=\{x_i(k)\,|\,i=1,2,\cdots\cdots,m\,;\,k=1,2,\cdots\cdots,n\},$$

The comparison sequence y is built by the data from the sea water quality standard,
y_j (k) represents the standard values of k at the level of j

$$y_j=\{y_j(k)\,|\,j=\text{I, II, III, IV}\,;\,k=1,2,\cdots\cdots,n\}.$$

Get the grey relational coefficient $L_{ij}(k)$,

$$L_{ij}(k)=\frac{min_j min_k \Delta ij(k)+\rho max_j max_k \Delta ij(k)}{\Delta ij(k)+\rho max_j max_k \Delta ij(k)} \tag{8}$$

$$\Delta ij(k)=|x_i(k)-y_j(k)| \tag{9}$$

Set ρ as the resolution coefficient, ρ=0.5[7].

3.2.3. Determine the weight of different indicators

The weight of each factor reflects its importance to the environment.

Set $\bar{S}_i(k) = \sum_{j=1}^{m} \bar{S}_i(k)/m; W_i = C_i(k)/\bar{S}_i(k); W_i(k) = W_i / \sum_{k=1}^{n} W_i$ (10) ~ (12)

Here $W_i(k)$ indicates the weight of i, which reflects the impact degree of this indicator on other indicators. Through which we can get the relative weight of the indicator.

3.2.4. Determine the grey relational coefficient
The grey relational correlation degree of the reference sequence and comparative sequence is obtained:

$$r_{ij} = \sum_{k=1}^{n} L_{ij}(k) W_i(k) \quad (13)$$

3.2.5. Evaluate the water quality
Calculate the water sample reference sequence and compare sequence correlation coefficient, determine the water quality evaluation results in accordance with the principle of correlation is the largest.

4. The Marine Water Quality Assessment
Taking the evaluation of marine water environmental quality in Qingdao as an example. The method of Bayesian and grey relational analysis was used to evaluate the quality of sea water in a sea area in Qingdao. For the convenience of comparison, the water quality monitoring data of 8 sampling points in the sea area of Qingdao are adopted, and COD, DO, inorganic-nitrogen, PO_4-P, oil, Cu, Zn, Pb, Cd are selected as evaluation factors. The measured values of sea water sampling points are found in the literature [8][9] (Table 1). The criteria for evaluation of each index are "People's Republic of China sea water quality standard" (GB3097-1997) (Table 2).

Table 1. Monitoring results of seawater quality in Qingdao (mg/L)

Monitoring station	COD	DO	inorganic-nitrogen	PO_4-P	Oil	Cu	Zn	Pb	Cd
M1	0.655	7.678	0.108	9.050	0.051	0.003	0.041	0.002	0.0002
M2	0.705	7.620	0.121	6.350	0.049	0.004	0.074	0.002	0.0002
M3	0.680	7.612	0.131	8.100	0.027	0.005	0.039	0.005	0.0003
M4	0.700	7.502	0.101	6.550	0.034	0.006	0.057	0.003	0.0002
M5	0.750	7.553	0.088	6.550	0.029	0.003	0.049	0.002	0.0003
M6	0.935	7.635	0.128	6.300	0.100	0.003	0.046	0.003	0.0002
M7	0.835	7.635	0.122	7.900	0.049	0.004	0.034	0.002	0.0001
M8	0.600	7.519	0.097	5.950	0.062	0.003	0.039	0.002	0.0002

Table 2. Sea water quality standards (mg/L)

Level	COD≤	DO>	inorganic-nitrogen≤	PO_4-P≤	Oil≤	Cu≤	Zn≤	Pb≤	Cd≤
I	2	6	0.20	15	0.05	0.005	0.020	0.001	0.001
II	3	5	0.30	30	0.05	0.010	0.050	0.005	0.005
III	4	4	0.40	30	0.30	0.050	0.100	0.010	0.010
IV	5	3	0.50	45	0.50	0.050	0.500	0.050	0.010

4.1. The evaluation results of Bayesian method

According to the calculation steps, the standard deviation of the index is calculated. (table3)

Table 3. Calculation results of standard deviation of various indexes

Index	I	II	III	IV
COD	0.73771	1.1066	1.4754	1.8443
DO	1.7213	1.4344	1.1476	0.86066
inorganic-nitrogen	0.073771	0.11066	0.14754	0.18443
PO_4-P	6.1237	12.247	12.247	18.371
Oil	0.048432	0.048432	0.29059	0.48432
Cu	0.0042821	0.0085642	0.042821	0.042821
Zn	0.026759	0.066898	0.1338	0.66898
Pb	0.0013718	0.006859	0.013718	0.06895
Cd	0.0006706	0.003353	0.006706	0.006706

The calculation results in Table 3 are replaced by the formula $t_i = \frac{xi-yij}{\sigma ij}$ for the standard. Then according to the formula (3), (4) and (5), the comprehensive posterior probability and evaluation result of the water quality are calculated. (Table 4)

Table 4. Evaluation results of Bayesian method

Monitoring station	I	II	III	IV	Evaluation grade
M1	0.43032	0.24111	0.17656	0.15201	I
M2	0.42859	0.24608	0.17598	0.14934	I
M3	0.43945	0.23905	0.17409	0.14742	I
M4	0.43277	0.24258	0.17525	0.14939	I
M5	0.42359	0.23884	0.18090	0.15667	I
M6	0.43878	0.24560	0.17091	0.14470	I
M7	0.43148	0.24159	0.17548	0.15145	I
M8	0.41773	0.24574	0.18003	0.15650	I

From the calculation results of table 4,$p_{\text{I}} > p_{\text{II}} > p_{\text{III}} > p_{\text{IV}}$.According to the principle of maximum correlation, the sea water of 8 monitoring points can be judged to be Ⅰ type of water. And the order of water quality of each monitoring point is M3>M6>M4>M7>M1>M2>M5>M8.

4.2. Evaluation Results of Grey Relational Analysis

The measured values in Table 1 are dimensionless treated in accordance with the formula (6) and (7), and the obtained sequence is used as a reference sequence. The concentration limit of each standard in Table 2 is the comparison sequence. According to the calculation steps, the results of grey relational degree calculation and the evaluation of water quality are obtained. (Table5)

Table 5. Evaluation results of grey relational analysis

Monitoring station	I	II	III	IV	Evaluation grade
M1	0.94347	0.91884	0.91751	0.89325	I
M2	0.94506	0.92196	0.91452	0.89013	I
M3	0.94396	0.91974	0.91712	0.89229	I
M4	0.94396	0.92149	0.91487	0.89041	I
M5	0.94496	0.92154	0.91481	0.89036	I
M6	0.94536	0.92210	0.91532	0.89091	I
M7	0.94423	0.92008	0.91726	0.89246	I
M8	0.94546	0.92232	0.91339	0.889127	I

From the calculation results of table 5,$r_{1\,\mathrm{I}} > r_{1\,\mathrm{II}} > r_{1\mathrm{III}} > r_{1\mathrm{IV}}$; $r_{2\,\mathrm{I}} > r_{2\,\mathrm{II}} > r_{2\mathrm{III}} > r_{2\mathrm{IV}}$; $r_{3\,\mathrm{I}} > r_{3\,\mathrm{II}} > r_{3\mathrm{III}} > r_{3\mathrm{IV}}$;$r_{4\,\mathrm{I}} > r_{4\,\mathrm{II}} > r_{4\mathrm{III}} > r_{4\mathrm{IV}}$. According to the principle of maximum correlation, the sea water of 8 monitoring points can be judged to be Ⅰ types of water. And the order of water quality of each monitoring point is M8>M6>M2>M5>M7>M4>M3>M1.

5. Comparison and Analysis

The evaluation results of the two methods are basically the same. All the water quality evaluation results of the monitoring points are I. It shows that Bayesian method based on normal distribution is suitable for evaluating the water quality of the sea water, and the result is accurate. However, there are differences in the sorting of water quality at different monitoring points by the two methods. Taking monitoring station M2 and M4 as an example, in the Bayesian method, the water quality of M4 is better than that of M2, while the result of grey relational analysis is the opposite. It can be found that these two methods are similar in thought, the difference is mainly due to the different ways of weight assignment.

6. Conclusion

Based on the Bayesian formula and the normal distribution principle, this article constructs the Bayesian water quality evaluation model. Taking seawater monitoring point in a sea area of Qingdao as an example, water quality evaluation is carried out, and the evaluation results are compared with those of grey correlation analysis, results show that, with modified Bayesian methods for sea water quality assessment is reasonable and feasible. Compared with the grey correlation analysis, the Bayesian method is more comprehensive and also applicable to small sample problems.

In this paper, taking into account the sampling error of the principle of normal distribution, the effects of the removal of the sampling error index value and the standard value of the cause, can be used to estimate objectively; in addition to many different calculation methods of large, complex calculation method, Bayesian principle is clear and the calculation is convenient and fast. In a word, Bayesian method is a feasible new method of marine water quality evaluation, which is worth popularizing.

7. Reference

[1] Changgen F,Yanzhou L.2008 Application of comprehensive evaluation method in environmental evaluation *Journal of Safety and Environment (*Vol.8 No.5*)* pp 112–115

[2] Lingling S, Shuqian W, Baohong S. 2017 Comprehensive assessment of water quality based on Bayesian Method & Normal Distribution Theory *Journal of Hebei University of Engineering (Natural Science Edition) (*Vol.34 No.4*)* pp 70–74

[3] Likun Y , Xinhua Z , Sen P , Xia L. 2016 Water quality assessment analysis by using combination of Bayesian and genetic algorithm approach in an urban lake, China *Ecological Modelling (*Vol.339*)* pp 77–88

[4] Ying Zhao, Ashish Sharma, Bellie Sivakumar, Lucy Marshall, Peng Wang, Jiping Jiang 2014 A Bayesian method for multi-pollution source water quality model and seasonal water quality management in river segments *Environmental Modelling & Software (Vol.57)* pp 216–226

[5] Yuan Y, Jian W. 2017 Advanced Grey Relational Analysis Method and Its Application in Water Quality Evaluation of the Lake-type Wetland *Journal of Landscape Research (*Vol.9 No.4*)* pp 81–83,87

[6] Jianmin B,Rui W, Xiaohan L. 2011 Comprehensive Assessment of Yitong River Quality Based on Bayesian Method *water saving irrigation (*No.5*)* pp 31–36

[7] Hui F, Yinglan S, Lei S. 2007 Application of grey correlation analysis in quality evaluation of marine environment *Transactions of Oceanology and Limnology (*No.3*)* pp 127–131

[8] Lin Z, Wenlin C, Yonggang J. 2007 Evaluation on seawater quality by fuzzy comprehensive evaluation method in Qingdao dumping area *Marine Environmental Science (*vol.26,No.1*)* pp 38–41

[9] Dongliang Z. 2012 *Study on the influence of long-term dumping on the marine environment of Qingdao ocean dumping area* (China: Ocean University of China)

IOP Conf. Series: Earth and Environmental Science **182** (2018) 012008 doi:10.1088/1755-1315/182/1/012008

Spatio-Temporal Change of Drought and Flood in northern Henan Province During 1961-2015 Based on Standardized Precipitation Evapotranspiration Index

Xin-Meng SHAN[1,2,a],Zhi-Guo LI[2,b,*] and Jia-Hong WEN[1,c]

1 Department of Geography, Shanghai Normal University, Shanghai 200234, China
2 Department of Surveying and Planning, Shangqiu Normal University, Shangqiu, Henan 476000, China
E-mail: [a] mxshan_123@163.com, [b] lizhiguo999999@163.com ,[c]jhwen@shnu.edu.cn,
[*]Corresponding author

Abstract. The spatial-temporal evolution of drought and flood events and their occurrence frequency in northern Henan Province, China are quantitatively analyzed during 1961~2015, based on temperature and precipitation data, by using Standardized Precipitation Evapotranspiration Index (SPEI) and indicators of drought and flood. The results show that there are significant differences in the periodicity and fluctuation of SPEI value at different time scales in the north of Henan Province during the study period. Firstly, at monthly scale, the monthly frequency of drought in the northern part of Henan Province shows a decreasing trend, but that of flood shows an increasing trend. Secondly, at seasonal scale, Autumn droughts have the highest occurrence frequency, followed by Summer droughts and Winter droughts. But for flood events, Summer floods have the highest frequency. Finally, as to the inter-annual variability of droughts and floods, serious droughts are occurred in northern Henan Province in the 1980s, 2000s and 2011-2015, but serious floods occur in the 1960s and 1970s.

1. Introduction

IPCC (Intergovernmental Panel on Climate Change) reports show that the temperature of global land surface has risen by 0.56-0.92℃ in recent 100 years. The global warming will increase the frequency and intensity of extreme weather disasters and has a great impact on agro-ecosystems[1]. Floods and droughts are among the most severe natural disasters in China, affecting a wide range of areas and causing serious losses. Since 1995, floods have accounted for 47% of all weather related disasters, affecting 2.3 billion people. The number of floods per year rises to an average of 171 in the period 2005-2014, up from an annual average of 127 in the previous decade[2]. In general, droughts last for months or even years [3-4]. Droughts are associated with agricultural failures, loss of livestock, drinking water supply shortages and outbreaks of epidemic diseases[5]. Therefore, quantitative study of temporal and spatial changes in floods and droughts explain its formation mechanism, which have great significance to disaster prevention and mitigation.

In recent years, Vicente-Serrano[6] and others propose the SPEI(Standardized Precipitation Evapotranspiration Index)[7-8]. Based on the SPI(Standardized Precipitation Index), SPEI further considers the amount of evapotranspiration which reflects the degree of drought and flood and becomes an important indicator system for measuring drought and flood [9-10]. In recent years, domestic researchers began to apply the SPEI index for scientific research. However, to our knowledge, few studies use SPEI to study and analyze the spatial and temporal variations of drought and flood systematically at small geographical scale such as the north of Henan Province.

This paper used the Standardized Precipitation Evapotranspiration Index (SPEI) to analyze the frequency of droughts and floods in northern Henan Province in recent 55 years. In the research, it contains the analysis of the successive drought and flood events and the average temperature and precipitation data of surface meteorological stations in the northern part of Henan Province from 1961 to 2015. Based on the monthly, seasonal, semi-annual and annual scale, the spatio-temporal variations of monthly drought and flood events and their occurrence frequency in northern Henan Province are studied in order to provide useful information for drought and flood disaster reduction decision-making and climate change in this region.

2. Study Region and Analysis Methods

2.1. Study Region

Northern Henan Province refers to the area north of the Yellow River in Henan Province, within the latitude of 34°47′N-36°20′N and longitude of 112°00′E-116°4′E [11]. It includes Xinxiang, Anyang, Jiaozuo, Puyang, Hebi five prefecture-level cities in total, and Jiyuan, the provincial-level cities [12] (Figure 1). The north of Henan Province has a population of 20 million people in 2015, covering an area of 28,000 km^2. The climate belongs to the warm temperate continental monsoon climate, the four seasons are distinct, the annual average temperature is 14℃, annual average rainfall amount is 650mm, mostly concentrated in Summer and Autumn[13].

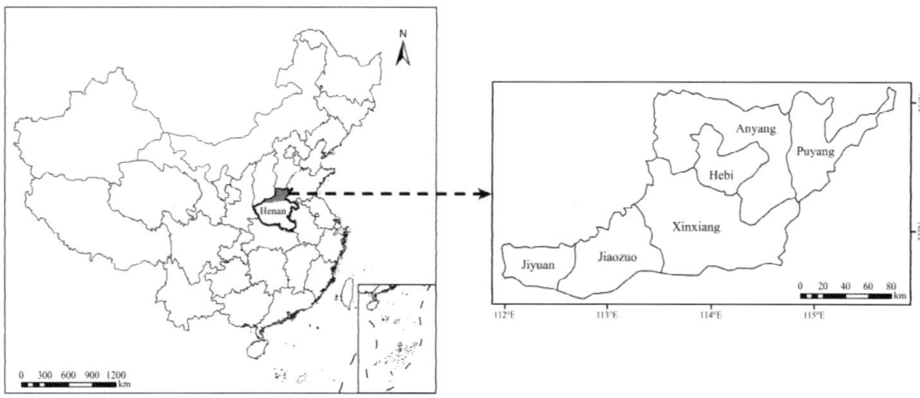

Figure 1. Distribution of weather station in the study

2.2. Data source

In this paper, monthly average air temperature and precipitation data of surface meteorological stations in Anyang and Xinxiang from 1961 to 2015 are used to calculate the standardized evapotranspiration index. The research scale of this paper is divided into four parts: monthly scale (SPEI-12), seasonal scale (SPEI-3), semi-annual scale (SPEI-6) and annual scale (SPEI-12). The monthly scale (SPEI-1) uses data from January to December each year, with a total of 660 sets of data. Special attention is payed to the seasonal scale (SPEI-3) in this work, except in 1961 February is used as the Spring season, March to May used as Spring for the remaining years, June to August as Summer, September to November as Autumn, December to February as Winter, a total of 220 data sets is used in analysis at season scale. The semi-annual scale (SPEI-6) is based on August as a proxy for the Summer-half year, February represents the Winter-half year, a total of 110 sets of data is used. The annual scale (SPEI-12) is based on the average of December each year, a total of 55 sets of data is used.

2.3. Analysis Methods

2.3.1. SPEI Index. The SPEI [9] is a standardized precipitation evapotranspiration index. The calculation methodis to characterize the degree of drought (flood) in a certain area by using the

deviation of the difference between precipitation and evapotranspiration from the mean state. The calculation steps of the index are as follows [6]:
Calculating the potential evapotranspiration (PET) with the Thornthwaite method:

$$PET_i = 16.0 \times \left(\frac{10T_i}{H} \right)^A \tag{1}$$

where: A is a constant; H is the annual calorific value; T_i is the average temperature of 30d. Calculation of constant A and annual heat index H. The formulae are expressed as follows:

$$A = 6.75 \times 10^{-7} H^3 - 7.71 \times 10^{-5} H^2 + 1.792 \times 10^{-2} H + 10^{-2} H + 0.49 \tag{2}$$

$$H_i = \left(\frac{T_i}{5} \right)^{1.514} \tag{3}$$

$$H = \sum_{i=1}^{12} H_i = \sum_{i=1}^{12} \frac{T_i}{5}^{1.514} \tag{4}$$

(2) Calculating the difference between monthly precipitation and evapotranspiration, the specific formula is as follows [9]:

$$D_i = P_i - PET_i \tag{5}$$

where: D_i is the difference between precipitation and evapotranspiration; P_i is the monthly precipitate-on; PET_i is the monthly evapotranspiration.

(3) D_i data sequence normalization. D_i is fitted by log-logistic probability distribution $F(x)$ to calculate the SPEI value corresponding to each D_i value.

when the cumulative probability $P \le 0.5$, the following formula is used to calculate $SPEI$.

$$W = \sqrt{-2\ln(P)} \tag{6}$$

$$SPEI = W - \frac{C_0 + C_1 W + C_2 W^2}{1 + d_1 W + d_2 W^2 + d_3 W^3} \tag{7}$$

Where: $d_1 = 1.432788$, $d_2 = 0.189269$, $d_3 = 0.001308$, $C_0 = 2.515517$, $C_1 = 0.802853$, $C_2 = 0.01032$.

When $P > 0.5$, using (8) to calculate $SPEI$.

$$SPEI = -\left(W - \frac{C_0 + C_1 W + C_2 W^2}{1 + d_1 W + d_2 W^2 + d_3 W^3} \right) \tag{8}$$

SPEI is characterized by multiple time scales. In this study, we mainly analyze the SPEI values at 1, 3, 6 and 12 months, which are respectively represented by SPEI-1, SPEI-3, SPEI-6 and SPEI-12 [7].

2.3.2. Drought and flood Evaluation Index. The spatio-temporal distribution characteristics of drought and flood are analyzed in northern Henan Province from 1961 to 2015, using the frequency and intensity of drought and flood. The specific cau-
lation steps are as follows [10]:
(1) drought (flood) frequency (P_i):
Calculating the frequency of occurrence of drought (flood) according to the number of occurrences of drought (flood) in a station. The calculation formula is as follows:

$$P_i = n / N \times 100\%$$

Where, N is the total number of years of study, n is the total number of years when a site is droughts (floods), $N = 55$. The frequency P_i is used to evaluate the frequency of droughts and floods in a station in the study area over the past 55 years.

Table 1. Standardized precipitation evapotranspiration index of drought and flood grade

IOP Conf. Series: Earth and Environmental Science **182** (2018) 012008 doi:10.1088/1755-1315/182/1/012008

Serial number	SPEI value	grade
1	$-1<SPEI\leq-0.5$	light drought
2	$-1.5<SPEI\leq-1$	middle drought
3	$-2<SPEI\leq-1.5$	heavy drought
4	$SPEI\leq-2$	very heavy drought
5	$-0.5<SPEI\leq0.5$	normal
6	$0.5<SPEI\leq1$	light flood
7	$1<SPEI\leq1.5$	middle flood
8	$1.5<SPEI\leq2$	heavy flood
9	$SPEI\geq2$	very heavy flood

(2) drought and flood intensity:
Drought and flood intensities are classified into four levels including light, middle, heavy and very heavy drought or flood. The SPEI value can reflect the drought and flood intensity at a certain weather station in a certain period of time. Table 1 shows the international drought and flood classification. However, because the occurrence of droughts and floods in northern Henan Province is less frequent, we combine the frequency of very heavy droughts and floods into the heavy droughts and floods frequency.

3. Results and Discussion

3.1. Analysis of Continuous Drought (Flood) events

The results of continuous drought and flood times and drought and flood duration in different sites in our study region from 1960 to 2015 are shown in table 2 and table 3. As far as drought events are concerned, the number of drought times at Xinxiang station are more than those at Anyang station. In addition, in 1960s, 1970s, 1980s, 1990s and 2000s, the number of consecutive drought in 2000s is the highest, which are five times higher and the longest duration is 16.5 months. In the case of flood events, the number of flood times at Anyang station are more than those at Xinxiang station. In addition, the number of consecutive flood with six times in 2000s is the highest, and the longest duration lasts 20.5 months.

Table 2. Continuous drought times and drought duration in different sites in northern Henan Province

Site/ Decade	1960s	1970s	1980s	1990s	2000s	2011-2015	Subtotal
Anyang	3T/12M	——	7T/23M	3T/10M	2T/7M	5T/16M	20T/68M
Xinxiang	2T/8M	1T/4M	2T/6M	4T/15M	8T/26M	3T/13M	20T/72M
Average annual	2.5T/10M	0.5T/2M	4.5T/14.5M	3.5T/12.5M	5T/16.5M	4T/14.5M	20T/70M

Note: T indicates the number of times, M indicates the number of months, — indicates no continuous drought (flood) event

Table 3. Continuous flood times and flood duration in different sites in northern Henan Province

Site/Decade	1960s	1970s	1980s	1990s	2000s	2011-2015	Subtotal
Anyang	4T/17M	3T/10M	2T/7M	2T/6M	3T/10M	2T/6M	16T/56M
Xinxiang	6T/24M	3T/12M	3T/11M	——	1T/5M	1T/3M	14T/55M
Average annual	5T/20.5M	3T/11M	2.5T/9M	1T/3M	2T/7.5M	1.5T/4.5M	15T/55.5M

Note: T, M meaning see table 2 note

3.2. Drought/Flood Features are Characterized by Fluctuation of SPEI Values at Different Time Scales

SPEI values have multi-time scale features, which can be divided into SPEI-1, SPEI-3, SPEI-6 and SPEI-12. The shorter the scale, the more elaborate the situation of drought and flood in northern Henan Province. The longer the scale, the more it can capture the trend of drought and flood in the north of Henan Province. There are obvious differences in the period and fluctuation of SPEI value in different time scales in the northern part of Henan Province (Figure 2). By analyzing the fluctuation of SPEI value at each time scale, we can draw the following conclusions: the fluctuation amplitude of SPEI is most obvious on the monthly scale; the volatility of the SPEI value on the seasonal scale is more gentler than the SPEI value on the monthly scale; Annual scale SPEI value fluctuations are most stable, reflecting the inter-annual variation of drought and flood.

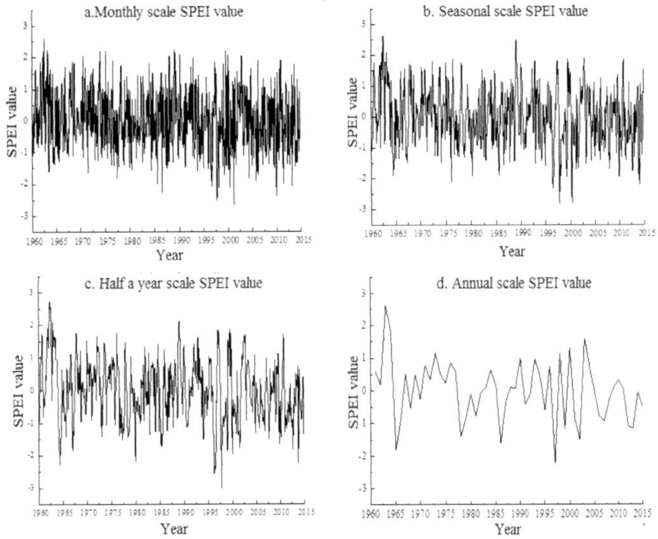

Figure 2. The annual SPEI values at different time scales in northern Henan province from 1960 to 2015

3.3. Drought and Flood Frequency Analysis

3.3.1. Drought and Flood Frequency Changes with Time. Figure 3a presents the inter-decadal drought frequency distribution of each site based on monthly values. The frequency of drought occurs at Xinxiang station in 2010s, and the frequency of drought is 48.33%.The frequency of drought at Xinxiang station in 1980s is 26.67%. The average of inter-decadal drought frequency is ranked as follows:2010s(45.00%)>1990s(37.50%)>2000s(37.09%)>198 0s(33.34%)>1960s(31.67%)>1970s (30.00%).

Figure 3b presents the inter-decadal flood frequency distribution of each site based on monthly values. It indicates that the maximum frequency of flood occurs at Anyang station in 2010s and Xinxiang station in 1960s, and the frequency of flood is 36.67%. The minimum frequency of flood occurs at Xinxiang station in 1990s, the frequency of flood is 23.33%. The average of inter-decadal flood frequency is ranked as follows: 1970s(35.83%)>1960s(34.59%)>1980s(34.17%)>2000s(32.50%)>201 0s(31.67%)>1990s (25.83%).

In general, the frequency of inter-decadal drought presents an increasing trend, and the decadal flood frequency presents a slight decreasing trend.

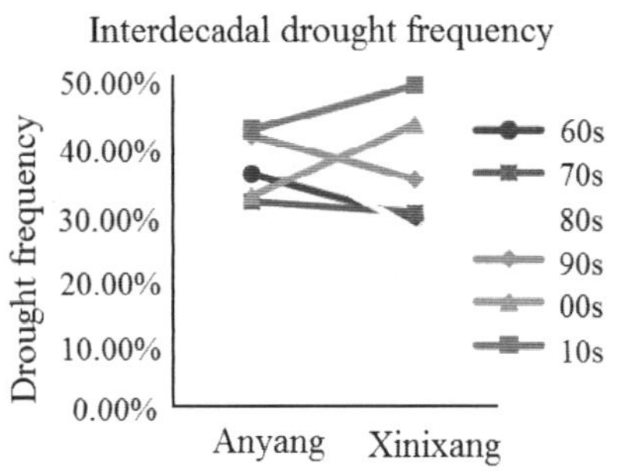

(3a)

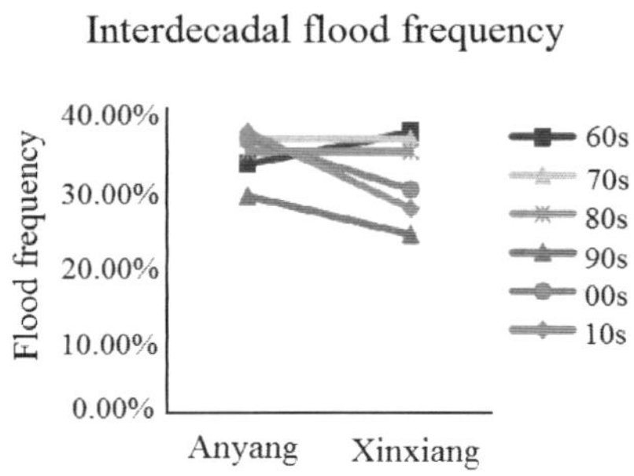

(3b)

Figure (3a). Interdecadal characteristics of light drought, moderate drought and heavy drought; **(3b).** Interdecadal characteristics of light flood, moderate flood and heavy flood.

3.3.2. Temporal Variation of Drought and Flood Frequency. From table 4, it can be seen that droughts occur most frequently in January and December, with the least frequency of drought in March and September; the most frequency of flood in April and May, and the lowest frequency of floods in March, September and November. Monthly drought at Anyang station occurs more frequently than that at Xinxiang station. Monthly flood at Xinxiang station occur more frequently than Anyang. The frequency of monthly drought in the northern part of Henan Province presents a decreasing trend, while the monthly frequency of floods presents an increasing trend.

Table 4. Drought and flood frequency distribution at month scale in northern Henan Province

	Anyang drought	Xinxiang drought	Anyang flood	Xinxiang flood
January	43.64%	41.82%	32.73%	32.73%
February	40.00%	30.91%	34.55%	27.27%
March	29.09%	30.91%	30.91%	30.91%
April	32.73%	36.36%	36.36%	36.36%
May	38.18%	34.55%	36.36%	32.73%
June	36.36%	30.91%	32.73%	32.73%
July	30.91%	27.27%	38.18%	29.09%
August	34.55%	30.91%	36.36%	30.91%
September	32.73%	34.55%	30.91%	29.09%
October	38.18%	36.36%	32.73%	30.91%
November	34.55%	36.36%	30.91%	30.91%
December	43.64%	34.55%	30.91%	32.73%
Average value	36.21%	33.79%	33.64%	31.36%

As shown in table 5, it is the droughts and floods frequency at seasonal scale in northern Henan Province. Generally speaking, at the seasonal scale, the frequency of Summer drought is the highest in the north of Henan Province, follows by the Autumn drought, while the Spring drought and Winter drought are relatively low. The frequency of Autumn flood is the highest in the northern part of Henan Province, followed by the Summer flood and Winter flood, and lowest in Spring flood.

Table 5. Drought and flood frequency distribution at season scale in northern Henan Province

	Anyang drought	Xinxiang drought	Anyang flood	Xinxiang flood
Spring	30.91%	27.27%	29.09%	23.64%
Summer	34.55%	38.18%	34.55%	34.55%
Autumn	34.55%	34.55%	38.18%	34.55%
Winter	29.09%	30.91%	34.55%	34.55%

Table 6. Decadal drought and flood frequency distribution in northern Henan Province

	Anyang drought	Xinxiang drought	Anyang flood	Xinxiang flood
1961-1970	35.00%	28.33%	32.50%	36.67%
1971-1980	30.83%	29.17%	35.83%	35.83%
1981-1990	40.00%	26.67%	34.17%	34.17%
1991-2000	40.83%	34.17%	28.33%	23.33%
2001-2010	31.67%	42.50%	35.83%	29.17%
2011-2015	41.67%	48.33%	36.67%	26.67%

The decadal drought and flood frequency distribution in northern Henan Province is shown in table 6. The frequency of droughts at the two stations decreases in the 1970s compared with the 1960s, but the frequency of floods increases. In the 1980s, the frequency of droughts at both sites indicates an increasing trend, and the frequency of floods demonstrates a slightly decreasing trend. However, in the 1990s, the frequency of droughts still illustrates an increasing trend, and the frequency of floods shows a dramatic decrease. In the 2000s, the frequency of droughts at two stations changes little and the frequency of floods displays an increasing trend. In particular, the frequency of droughts in 2011-2015 is higher than all previous years.

It has increased by 15 percentage compared with previous decade, while the frequency of floods has only dropped by 2 percentage. In short, the frequency of drought in northern Henan Province shows an increasing trend at decadal scale, and the frequency of floods shows a decreasing trend.

3.3.3. Spatial Variation of Drought and Flood Frequency. Based on SPEI-1 drought and flood statistics at each site, the frequency of drought and flood in Henan Province over the past 55 years and the distribution of drought and flood frequencies at all levels are shown in Figure 4(a) and 4(b).

For Henan Province, the maximum frequency of drought occurred at Xinyang and Xixia station(36.36%), which was higher than the average drought frequency in the whole province (32.34%); the minimum occurred at Zhengzhou station (25.45%). The frequency of drought in northern Henan Province gradually decreased to the east, and the minimum value of drought occurred at Anyang station (29.09%). The frequency of light drought in northern Henan Province decreased from west to east, the minimum frequency of light drought was Anyang station (10.91%), the maximum frequency was located in Xinxiang (16.36%), and the frequency of middle drought is around 10.91%; the heavy drought frequency is stable at 7.27%.

In the past 55 years, the maximum flood frequency in Henan Province is located in Nanyang (47.27%), which is higher than the average flood frequency in the whole province (34.63%). The frequency of flooding in northern Henan Province remained unchanged at 32.73%. For northern Henan Province, the maximum frequency is Xinxiang station , which is 21.82% and the minimum is Anyang station, which is 16.36%; the median frequency is 9.09% at Anyang station, and the minimum is 3.64% at Xinxiang station; the frequency of heavy flood at both sites is the same, which is 7.27%.

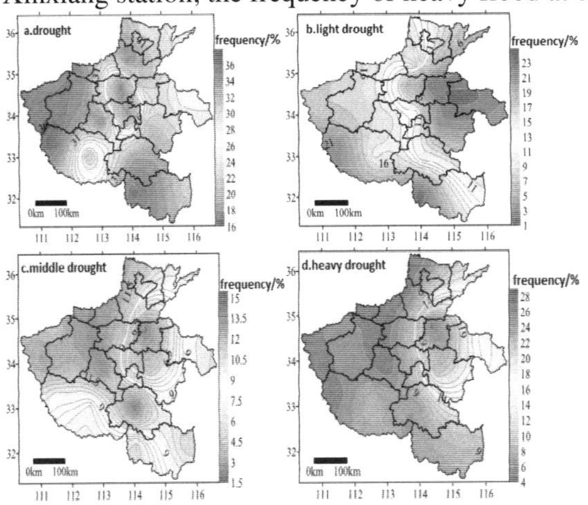

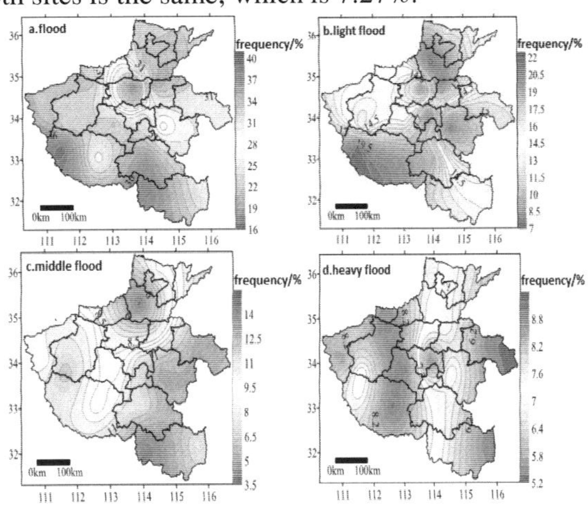

Figure 4(a). Spatial variation of drought frequency in Henan Province during 1961-2015;
4(b). Spatial variation of flood frequency in Henan Province during 1961-2015

4. Conclusions

By analyzing the spatio-temporal variations of drought and flood at different time scales and spatial ranges from 1961 to 2015 in northern Henan Province, we can draw the following conclusions:
(1) The fluctuation amplitude of SPEI value is inversely proportional to the time scale and the fluctuation period of SPEI value is proportional to the time scale. In the distribution trend of average SPEI over different time scales, SPEI-12 fluctuates the most, SPEI-6 follows by SPEI-3, and SPEI-1 fluctuates the least.
(2) The average number of consecutive drought occurrences in the north of Henan Province is 3.5 times, which Xinxiang is higher than this average from 1961 to 2015. The average number of flood occurs in northern Henan Province is 3.7 times, which Anyang is lower than this mean value. The number of consecutive droughts, floods and cumulative durations of all stations in the northern part of Henan Province tends to decrease with time.
(3) The frequency of light and middle drought is higher than heavy drought in northern Henan Province over 55 years; the frequency of light and middle flood is higher than heavy flood. From the perspective of time evolution, the frequency of middle drought in the north of Henan Province in the recent decade (2010s) is higher than that in 1960s, and the frequency of middle flood in 1960s and 1990s is higher than that in 1960s. Our findings reflect drought and flood disasters in northern Henan Province, which show an increasing trend under the background of global warming.
(4) The drought frequency at Anyang station is higher than that at Xinxiang station, the flood frequency at Xinxiang station is higher than that at Anyang. The specific geographic location and climatic conditions in the north of Henan Province have developed their obvious characteristics of spatio-temporal change of drought and flood disasters. It is different from other areas of Henan Province in term of seasonal, regional and inter-annual variations.
(5) The frequency of drought in northern Henan Province gradually decreases to the east, and the minimum value of drought occurs at Anyang station(29.09%). The frequency of flooding in the northern part of Henan Province remained unchanged at 32.73%. For northern Henan Province, the maximum frequency is Xinxiang station, which is 21.82% and the minimum is Anyang station, which is 16.36%.

5. References

[1] Intergovernmental Panel on Climate Change. Climate Change 2013: *The Physical Science Basis.* Contribution of Working Group I to the Fifth Assessment Report of the Intergovernmental Panel on Climate Change; Cambridge University Press: New York, NY, USA, 2013.
[2] Human Cost of Weather Related Disasters (1995-2015), Centre for Research on the Epidemiology of Disasters, *UN Office for Disaster Risk Reduction* Published on 23 NOV 2015.
[3] D B Lobell, A Sibley, J I Ortiz-Monasterio. Extreme heat effects on wheat senescence in India, *Nature Climate Change.* 2(3)(2012) 186-189.
[4] D.H. Qin, Climate Change: Regional Response and Prevention and Mitigation of Disaster, Beijing: *Science Press.* 2009 17-19.
[5] The human cost of natural disasters 2015: a global perspective Centre for Research on the Epidemiology of Disasters, *UN Office for Disaster Risk Reduction* Published on 06 Mar 2015.
[6] Vicente-Serrano S M, Beguería S, López-Moreno J I, A multiscalar drought index sensitive to glo-bal warming: the standardized precipitation evapotranspiration index, *Journal of Climate.* 2010, 23(7):1696-1718.
[7] L.Y. Xu, H. M. Wang, Q. C. Duan and J. F. Ma, Spatial and Temporal Characteristics of Summer Maize Growth Season in Yunnan Province Based on SPEI, *Resources Science.* 05(2013) 1024-1034.
[8] X.Y. Zhu, Drought evolution characteristics of eastern Fujian over the past 50 years Based on SPEI, *Journal of Natural Disasters.* 04(2015) 128-137.

[9] B.L. Shi, X.Y. Zhu, Y.C. Hu and Y.Y. Yang, Spatial and Temporal Variations of Drought in He-Nan Province in Recent 53 Years Based on SPEI Index, *Geographical Research.* 08(2015)1547-1558.
[10] K. Liu, D.B. Jiang, Based on two potential evapotranspiration algorithms SPEI analysis of the C-hange of Wet and Dry in China, *Chinese Journal of Atmospheric Science*. 2015,39(01):23-36.
[11] J. Zhao, Analysis on the Risk of Drought Disaster in North Henan under the Background of Climate Change. *Northeast Normal University,* 2012.
[12] L. Tang, J.Y. Luan and R.H. Liu, Energy Saving Technology of Traditional Courtyard Houses in Henan North , *Building Science*. 2012, 28(06):10-13+23.
[13] J. Zhao, J.Q. Zhang, D.H. Yan, Z.J. Tong, X.P. Liu, Disaster risk zoning in northern Henan based on grid GIS [J], *Journal of Catastrophology*. 2012,27 (01): 55-58.

Acknowledgments

This work was supported by the National Natural Science Foundation of China (Grant No. 41671072), Foundation for University Young Key Teacher by the Henan Educational Committee (Grant No. 2016GGJS-130), Key Scientific Research Project of Universities in Henan Province (Grant No. 15A170011). Meteorological data of Henan Province were kindly provided by the China Meteorological Data Sharing Service System.

IOP Conf. Series: Earth and Environmental Science **182** (2018) 012009 doi:10.1088/1755-1315/182/1/012009

Insecticide Usage in Lotus-Fish Farming and Its Impact on Fish Culture and Grower Health

S. Bumroongsook

Department of Plant Production Technology, Faculty of Agricultural Technology, King Mongkut's Institute of Technology Ladkrabang, Bangkok 10520, Thailand
E-mail: suvarin.bu@kmitl.ac.th

Abstract. Insect pest is a major problem of lotus flower production for local markets and export. Lotus growers inappropriately use high quantity of toxic insecticides to control insect pest. Insecticides are applied all over the lotus farming and the chemical substances are not limited to lotus flowers but they leach down to the water reservoir affects fish and other nontarget living organisms in water and the environment. Therefore, the survey of insecticides usage and related information for lotus-fish farming in 3 different provinces: Chachoengsao, Suphanburi and Nakhonpathom was conducted. The result showed most growers had an elementary school certificate while the rest held a secondary school certificate. The important insect pest of lotus are thrips (*Frankliniella schultzei*) and common cutworm (*Spodoptera litura*). Lotus growers preferred organophosphate insecticides the most, followed by the avermectin group. Most growers from Chachoengsao(89.9%), Nakhonpathom(84.1%) and Suphanburi(77.4%) had insecticide application regularly. Decontamination of insecticidal residue on lotus flowers after harvest could be done by using flower dipping in water before distribution. Concerning over fish culture in lotus-fish farming, growers used the insecticide that had least effect on fish. However, the chemical application might cause the slow growth of fish. None of fish kill occurred due to insecticide application for insect control in lotus-fish farming. Repeated exposure to insecticide during application causes health problems to the growers.

1. Introduction

Lotus plant (*Nelumbo nucifera* Geartn) is a potential economic plant, since it can be used in a variety of floral ornamental plants, food and pharmacueticals. Lotus business is a national and international business, especially *N. nucifera* is the most commercially lotus. Presently, China, Australia and Vietnam has classified lotus as one of an important plant for an export industry.

Insect pest has a huge impact on lotus production. Lotus growers are dealt with insect pest during preharvest and postharvest. *Frankliniella schultzei* (Trybom) causes damage to lotus flower while *Scirtothrips darsalis* Hood is found infested in both flowers and young leaves. These thrips are small insects and their cryptic hebavior make them very difficult to detect and control [1]. Common cutworm (*Spodoptera litura* Fabricius) is a polyphagous insects which can destroy the lotus farming within a short period of time during peak season. The common cutworm is a heavy leaf feeder [2]. The cutworm larvae display resistance to pyrethroids, organophosphate, insect growth regulators and Spinosad, emamectin benzoate, indoxacarb and abamectin [3]. Thus, the chemical control of *S. litura* is increasing difficult and they have resistance to most common insecticides [4]. The insecticide resistance of cutworm populations was due to multiple resistance mechanisms, increased detoxification process and unresponsive acetyl-cholinesterase [5]. Therefore, lotus growers inappropriately use high quantity of toxic insecticides to control these insect pest and often could not effectively eliminate the insect pest.

The main problem in lotus production is the high cost of insect pest control. In addition, they spray insecticides all over the crops to prevent production loss. The chemical substances are not limited to lotus plants but they run off to the water source. Therefore, insecticide application affects fish and other nontarget organisms living in water and its environment. Theoretically, the insecticide spray should be lethal to the target insects, not other organism and human [6]. The decision making for judicious insecticide application is necessary for the protection of agricultural crops [7]. Insecticides are used for crop protection and to suppress insect pest population which its outbreak cause production loss and lower the species diversity in the areas. Nevertheless, these insecticides especially broad sprectrum pesticides had a serious negative impact on the environment.

Less than 2 % of sprayed insecticides reach target species, because the application are sprayed across the whole agricultural areas. Agrochemical substances used for pest control is hazardous to fish. When insecticides entry into aquatic systems via run off or spray drift, it had effect on nontarget organisms. Pesticide-associated fish kills are reported the incidence in the United States. These death of fish sometime involved thousands of fishes, as well as frogs, turtles, mussels, water birds, and other organisms. Pesticide contamination in water reservoirs is one of many factors contributing to the decline of fish [8]. Synthetic pyrethroids are the most toxic group of insecticides to fish and other aquatic invertebrate fuana. However, they rarely cause fish kills because of its physical and chemical properties: absorbed to bottom muds, a very short half-life in the environment, rapidly decomposition in sunlight, and lower application rates[8]. In the aspect of cytopathological studies, pesticides caused disintegration of metabolism and sometimes fish kill [9]. It effected mainly in the liver, blood vessels, kidneys, and gills. Cytoplasmic granularity and liver cell mass contraction might occur. Methoxychlor caused pathological changes of large blood vessels in bluegill by [10]. Organochlorine insecticides induced abortions in mosquito fish [11]. Continuous insecticide application on aqutic plants may pose risk to human health via consumption of contaminated fish. The adverse effect on fish and other organisms is related to the toxicity of insecticides, dosage and environmental factors [12].

Fipronil is a non selective insecticide of the phenylpyrazole class while acetamiprid is a widely used third generation pesticide of the neonicotinoid group. Acetamiprid is for sucking insect pests treatment especially thrips, plant hoppers and aphids in agricultural farming [13] and can be removed from waste water by electrocoagulation [14]. As lotus cut f lowers require the use of insecticides to control thrips, which cause damage to production and marketability value. Therefore, exported cut flowers receive insecticide treatment applications prior to shipment. Maximum residue limits (MRL) for flowers is not available, unlike the food products. There is no limitation on the use of pesticides before and after harvest for cut flowers [15]. Some pesticides used are persistent and easily be absorbed through skin contact. Handling the contaminated flowers for a long period of time daily, the florists can be exposed to pesticide deposits and possibly cause illness. Health problems for florist workers included contact allergies, dermatitis and skin effects [16-17]. Neurologic expression [15,18-22]. The flower decontamination can alleviate these problems.

Few studies has been done on the insecticide used in lotus-fish farming of lotus growers in commercial lotus growing area. Therefore, the objective of the present research is conduct a survey and collect data on the education background, insecticide usage and application including the effect on fish culture and grower health from 3 different provinces in Thailand: Chachoengsao, Nakhonpathom and Suphanburi. In addition, the decontamination of insecticide after harvest by water dipping method is assessed.

2. Materials and Methods

2.1. Insecticide usage and its impact on fish culture and grower health

The questionnaire survey is used to collect information from lotus grower about education background, pest problem knowledge, guidance for insecticide usage, its impact on fish culture and illness of growers due to insecticide application.

Respondents were 90 lotus growers from 3 provinces: Chachoengsao, Nakhonpathom and Suphanburi(30 respondents/province). Data were collected using a pretested questionnaire via the face-to-face interviews.

2.2. Decontamination of fipronil and acetramiprid on lotus flower by water dipping

Two insecticides (fipronil and acetramiprid) are choosen for the experiment. The experimental design was completely randomized with 3 treatment and 5 replication as follows:
1. Lotus flower dipping in the test insecticide for a couple minutes and residue analysis on the sample was conducted.
2. Lotus flower dipping in the test insecticide for a couple minutes, 7 hours after the treatment and residue analysis on the sample was conducted.
3. Lotus flower dipping in the test insecticide for a couple minutes, 7 hours after the treatment, then water dipping before residue analysis on the sample.

The analysis of insecticidal residues were performed followed Steinwandter [23]. Universal 5 minutes online method for insecticide residues is based on the Fresenius Journal of Analytical Chemistry no 1155. The GC-MS was used to perform the analysis.

2.3. Data Collection and Statistical Analysis

A personal interview survey is conducted to explore the responses of lotus growers and gather more and deeper information. During the interview, insect pest is collected for further identification and insecticide containers are observed and recorded the trade names and common names. The descriptive statistics analysis was performed. ANOVA analysis was used to determine the mean difference among treatments on decontamination by flower dipping in water

3. Results and Discussion

3.1. Insecticide usage

Results from the survey showed that most lotus grower had elementary school certificate (100% in Chachoengsao, 76.6% from Nakhonpathom and 94.1% from Suphanburi), the rest held secondary school certificate (Table 1). All lotus growers indicated that they faced insect problems for lotus production (Table 2). These growers reported that there were two destructive insect pest caused heavy damage to lotus plants. One is the commom cutworm which is known as a heavy leaf feeder. The other insect pest is the common blossom thrips (*Frankliniella schultzei*) which infested and caused black spots on lotus flowers. The common cutworm are resistance to most common insecticides [4]. Thrips control is difficult due to its small body size and the cryptic hebavior [1]. The growers from these three provinces solely depend on chemical control. They use common insecticides (abamectin, carbaryl, chlorpyrifos, cypermethrin, dimethoate, endosulfan, fenobucarb, methamidophos, methyl parathion, monocrotophos) which 4 of these insecticides were banned from use by the government (Table 3). These growers preferred organophosphate insecticide group the most and followed by the avermectin, pyrethroid and carbamate, respectively (Table 4). Most growers controlled insect pests using calendar schedules for treatments (Table 5). Growers received information of insecticide knowledge from grower companion, extension agents, self trial and error, sale representatives and agrochemical retailers (Table 6). Thai farmers relied on chemical control for crop protection and pesticide usage increased up to four fold in the past decade [24].

3.2. Decontamination of fipronil and acetamiprid on lotus flower by water dipping

Fipronil and acetamiprid are used for thrips control in lotus flowers. The decontamination using flower dipping method in water could reduce acetamiprid residue to 0.7667 mg/kg but not for fipronil (Table 7). MRL is not required for flowers, thus no limitation on the use of pesticides for cut flowers [15]. Some pesticides used are persistent and harmful to florists and professional workers who exposed and handling the products [16-17]. Removal insecticide residue can be done in a simple way.

3.3. Effect of insecticide usage on fish culture

Lotus growers released various juvenile fish such as *Hypophthalmichthys molitrix*, *H.nobilis*, *Trichogaster trichopteru*, *Barbodes gonionotus*, *Oreochromis niloticus*. Two types of fish culture (semi food feeding and natural food) are located in lotus-fish farming in Chachoengsao, Nakhonpathom and Suphanburi(Table 8). Freshwater fish are benefit as insect predators which help to lower aquatic insects.

Table 1. Education background of lotus growers in selected provinces

Education level	Percentage of growers(n=30)		
	Chachoengsao	Nakhonpathom	Suphanburi
Elementary school	100.0	76.6	94.1
Secondary school	0.0	23.4	5.9
Vocational school	0.0	0.0	0.0
University	0.0	0.0	0.0
Total	100.0	100.0	100.0

Table 2. Insect pest problem of lotus production

Pest of lotus	Percentage of growers(n=30)		
	Chachoengsao	Nakhonpathom	Suphanburi
Thrips	47.6	45.0	39.4
Common cutworm	52.4	42.3	40.0
Hairy caterpillar	0.0	5.6	14.1
Red mite	0.0	7.1	6.5
Total	100.0	100.0	100.0

Table 3. Different insecticide used by growers from targeted provinces

Common names	Banned insecticide	Province		
		Chachoengsao	Nakhonpathom	Suphanburi
abamectin		/	/	/
carbaryl		/		
chlorpyrifos		/	/	
cypermethrin		/	/	/
dimethoate	/	/		
endosulfan	/	/	/	/
fenobucarb			/	/
methamidophos	/	/	/	/
methyl parathion		/	/	/
monocrotophos	/		/	

Table 4. Insecticide groups used by lotus growers

Insecticide group	Percentage of growers(n=30)		
	Chachoengsao	Nakhonpathom	Suphanburi
organochloride	11.1	0.0	1.9
organophosphate	46.7	48.4	39.4
carbamate	11.1	1.6	0.0
pyrethroid	11.1	3.2	21.2
Avermectins	20.0	46.8	37.5
Total	100.0	100.0	100.0

Table 5. Time table for insecticide application by growers(in percentage)

Time frame	Percentage of growers(n=30)		
	Chachoengsao	Nakhonpathom	Suphanburi
Calendar spray	89.9	84.1	77.4
Insect infestation	11.1	15.9	22.6
recommendation	0.0	0.0	0.0

Total	100.0	100.0	100.0

Table 6. Source of insecticide knowledge

Source of knowledge	Percentage of growers(n=30)		
	Chachoengsao	Nakhonpathom	Suphanburi
Grower companion	0.0	39.9	56.1
Extension agent	0.0	0.0	0.0
Trial and error	47.2	19.1	2.8
Sale representative	0.0	5.4	5.9
Chemical retailers	52.8	35.6	35.2
Total	100.0	100.0	100.0

Table 7. Decontamination of insecticidal residue on lotus flowers

treatment	Fipronil[1] (mg/kg)	Acetamiprid[1](mg/kg)
0 hr	0.1933a	3.4067a
7 hrs	0.0933a	2.2133ab
7 hrs and water dipping	0.0900a	0.7667b

[1]Means followed by same letter do not significantly differ (P=.05, DMRT)

Some growers used light traps to attracted insects for fish food. Only 4 growers indicated insecticide spray had effect of fish growth(Table 9). stated that Some insecticides had negative effects on the growth and reproduction of fish [25-26]. Nile tilapia raised alone in pond had significantly growth more than co-cultured with lotus [27]. Epiphytic algae which are fish food used lotus plant as substrate [27-28]. The lotus-fish farming can recycle nutrient effectively [27].

3.4. Effect of insecticide application on growers

Lotus growers prefer power sprayers more than knapsack sprayers to spray pesticides over the crop plants on a small boat or spray and walk through the lotus farm. Chemical drift always occurs during the spray application [29]. All of them do not use protective clothes during the spray. Some of them use a small piece of clothes to cover their mouths and noses during the application. Although, they try to clean up themselves right away after the spray, insecticides are toxic and potentially hazardous to farmers and other organisms, including the environment. Farmers who regularly spray insecticides to protect their crops, insecticide droplets come in contact with them. Insecticide poisoning has been reported by lotus growers (10% in Chachoengsao, 19.9% in Nakhonpathom and 24.3% in Suphanburi)(Table 10). The common symptoms are often headache and/or dizziness, excessive salivation, difficult breathing, sweating and skin irritation. However, most growers are not affected by insecticide handing and practice. Both organophosphates and carbamates could lead to eye tearing, unclear vision, enormous saliva, more sweating, coughing and regurgitation. Pyrethrins group can cause similar symptoms including infrequent difficulty breathing [6]. Illness due to organophosphoorus insecticide contact did not differ significantly between spraying and nonspraying periods [30]. Health factors related to insecticide exposure, poor handing practices, and contamination in environment that could cause poisoning cases among farmers [31]. Long time exposure to combination of pesticides may result in unknown adverse health effects , therefore new low risk and cleaner agriculture practice should be urgent implemented for human health [32].

Table 8. Type of lotus-fish farming

Type of fish culture	Percentage of growers(n=30)		
	Chachoengsao	Nakhonpathom	Suphanburi
Semi food feeding	55.5	43.4	47.1
Natural food	44..5	56.6	52.9
Total	100.0	100.0	100.0

Table 9. Effect of insecticide application on fish

Fish symtom	province		
	Chachoengsao	Nakhonpathom	Suphanburi
Slow growth		/	/
Fish kill			

Table 10. Characteristics of illnesses associated with insecticide application

Symtoms	Percentage of growers(n=30)		
	Chachoengsao	Nakhonpathom	Suphanburi
Headache and/or dizziness	6.7	3.3	6.7
Bowel movements and urination	0.0	0.0	0.0
Muscles twitch	0.0	0.0	0.0
Excessive salivation	3.3	10.0	3.3
Difficult breathing	0.0	3.3	6.7
Sweating	0.0	3.3	3.3
Skin irritation	0.0	0.0	3.3
None	90.0	80.1	76.7
Total	100.0	100.0	100.0

4. Conclusions

Information on insect pest problems and control management from government sectors. is not available for lotus growers. Knowledge and practice of insecticide usage is from trial and error, chemical merchants and grower companion. All lotus growers primarily use insecticides to suppress insect population. Some insecticides banned or prohibited for use in agricultural areas by the government are still used in some areas. A simple method using flower dipping in water would decrease insecticidal contaminants. Chemical control is unavoidable for lotus production. Rational decision making process to determine the use and select insecticides wisely will minimize the impact on non targeted organism and environment. Lotus growers'awareness and safe use of insecticides should be implemented. The coupling of lotus farming with fish production could be established in commercial lotus growing areas because it increase the growers'income.

5. References

[1] Tyagi K, Kumar V, Singha D and Chakraborty R. 2015 Morphological and DNA Barcoding Evidence for Invasive Pest Thrips, *Thrips parvispinus* (Thripidae: Thysanoptera), Newly Recorded From India *J Insect Sci.* 15(1) 105

[2] Ahmad M, Ghaffar A and Rafiq M 2013 Host plants of leaf worm, *Spodoptera litura* (Fabricius) (Lepidoptera: noctuidae) in Pakistan *Asian J Agri Biol,* 1(1) 23-28.

[3] Saleem M, Hussain D, Ghouse G, Abbas, M and Fisher SW 2016 Monitoring of insecticide resistance in *Spodoptera litura*(Lepidoptera: Noctuidae) from four districts of Punjab, Pakistan to conventional and new chemistry insecticides *Crop Prot.* 79 177-184.

[4] Tong H, Su Q, Zhou X and Bai L 2004 Field resistance of *Spodoptera litura* (Lepidoptera: Noctuidae) to organophosphates, pyrethroids, carbamates and four newer chemistry insecticides in Hunan, China *J Pest Sci* 86(3)599-609.

[5] Yonggyun K., Cho JR, Lee J, Kang S, Han S., Hong KJ, Kim HS, Yoo JK, Lee JO 1998 Insecticide Resistance in the Tobacco Cutworm, *Spodoptera litura* (Fabricius) (Lepidoptera: Noctuiae) *J Asia-Pac Entomol* 1 (1) 115-122.

[6] Aktar MW, Sengupta D and Chowdhury A 2009 Impact of pesticides use in agriculture: their benefits and hazards *Interdiscip Toxicol.* 2(1)1-12.

[7] Mahmood Q, M. Bilal Mand Jan S 2014 Herbicides, Pesticides, and Plant Tolerance: An Overview *Biological Technique* 1:423–448.

[8] Helfrich LA 2009 *Pesticides and Aquatic Animals: A Guide to Reducing Impacts on Aquatic Systems.* Virginia Cooperative Extension 420-013.

[9] Murthy KS, Kiran BR and Venkateshwarluk M 2013 A review on toxicity of pesticides in Fish *International Journal of Open Scientific Research* 1(1):15-36.

[10] Kennedy HD, Eller LL and Walsh DF 1970 Chronic Effects of Methoxychlor on Bluegills and Aquatic Invertebrates *Tech. Pap. Bureau Fish. Wildl.* 53:3-18.

[11] Boyd CE 1964 Insecticides cause mosquitofish to abort *Prog. Fish-Cult.* 26 138.

[12] Damalas CA and Eleftherohorinos IG 2011 Pesticide exposure, safety issues, and risk assessment indicators *Int J Environ Res Public Health* 8(5) 1402–1419

[13] SinghTB, Mukhopadhayay SK, Sar TK and Ganguly S 2012 Induced acetamiprid toxicity in mice: A review *J Drug Metab Toxicol* 3:6

[14] Johna S, Solomanb PA and Fasnabi PA 2015 Study on removal of Acetamiprid from wastewater by Electrocoagulation *Procedia Technology* 24 (2016) 619 – 630.

[15] Toumi K, Vleminckx C, van Loco J and Schiffers B 2016 Pesticide residues on three cut flower species and potential exposure of florists in Belgium *Int J Environ Res Public Health* 13(10) 943.

[16] Das R, Steege A, Baron S, Beckman J and Harrison R 2001 Pesticide-related illness among migrant farm workers in the United States *Int. J. Occup. Environ. Health* 7:303–312.

[17] Penagos H, Ruepert C, Partanen T, Wesseling C 2004 Pesticide patch test series for the assessment of allergic contact dermatitis among banana plantation workers in panama. *Dermat. Contact Atop. Occup. Drug* 15:137–145.

[18] Faraha TM, Abdelrasoul GM, Amr MM, Shebl MM, Farahat FM and Anger WK 2003 Neurobehavioural effects among workers occupationally exposed to organophosphorous pesticides. *Occup. Environ.* 60 279–286.

[19] Alavanja MC, Hoppin JA and Kamel F 2004 Health Effects of Chronic Pesticide Exposure: Cancer and Neurotoxicity *Annu. Rev. Public Health* 25 155–197.

[20] Lu JL 2005 Risk factors to pesticide exposure and associated health symptoms among cut-flower farmers *Int. J. Environ. Health Res.* 15 161–170.

[21] Wesseling C, De Joode BVW, Keifer M, London L, Mergler D and Stallones L 2009 Symptoms of psychological distress and suicidal ideation among banana workers with a history of poisoning by organophosphate or n-methyl carbamate pesticides. *Occup. Environ. Med.* 67 778–784.

[22] Baldi I, Gruber A, Rondeau V, Lebailly P, Brochard P and Fabrigoule C 2011 Neurobehavioral effects of long-term exposure to pesticides: Results from the 4-year follow-up of the PHYTONER Study. *Occup. Environ. Med.* 68 108–115.

[23] Steinwandter H 1985 Analytical Methods for Pesticide and Plant Growth Regulators vol XVII, ed J Sherma(New York: Academic Press INC)

[24] Panuwet P, Siriwong W, Prapamontol T, Ryan PB, Fiedler N, Robson MG and Barr DB. 2012. Agricultural pesticide management in Thailand: situation and population health risk *Environ Sci Policy* 17 72-81.

[25] Hanson R, Dodoo DK, Essumang DK, Blay J Jr, Yankson K. 2007 The effect of some selected pesticides on the growth and reproduction of fresh water Oreochromis niloticus, Chrysicthys nigrodigitatus and Clarias gariepinus. *Bull Environ Contam Toxicol.* 79(5) 544-7.

[26] Sabra FS and Mehana. E-S E-D 2015 Pesticides Toxicity in Fish with Particular Reference to Insecticides *Asian Journal of Agriculture and Food Sciences* 3 (1)40-60.

[27] Yang Y, Lin CK and Diana JS 2002 Lotus-Fish Culture in Ponds: Recycling of Pond Mud Nutrients In: K. McElwee, K. Lewis, M. Nidiffer, and P. Buitrago (Editors), Nineteenth Annual Technical Report. Pond Dynamics/Aquaculture CRSP, Oregon State University, Corvallis, Oregon

[28] Shrestha MK and Knud-Hansen CF 1994 Increasing attached microorganism biomass as a management strategy for Nile tilapia (*Oreochromis niloticus*) production *Aquacult. Eng.* 13:101–108.

[29] Glotfelty J and Schomburg J 1989 Volatilization of pesticides from soil in Reactions and Movements of organic chemicals in soil. In: Sawhney BL, Brown K, editors. Madison, WI: Soil Science Society of America Special Pub

[30] Ngowi AV, Maeda DN, Partanen TJ, Sanga MP and Mbise G 2001 Acute health effects of organophosphorus pesticides on Tanzanian small-scale coffee growers *J Expo Anal Environ Epidemiol.* 11(4):335-9.
[31] Lu JL, Cosca KZ and Del Mundo J 2010 Trends of Pesticide Exposure and Related Cases in the Philippines *J Rural Med.* 5(2):153-64.
[32] Nicolopoulou-Stamati P, Maipas S, Kotampasi C, Stamatis P and Hens L 2016 Chemical Pesticides and Human Health: The Urgent Need for a New Concept in Agriculture *Front Public Health.* 18;4:148.

Acknowledgement
This presentation followship is supported by a grant from Faculty of Agricultural Technology, King Mongkut's Institute of Technology Ladkrabang.

IOP Conf. Series: Earth and Environmental Science **182** (2018) 012010 doi:10.1088/1755-1315/182/1/012010

A Preliminary Study on Application of MBR+NF/RO (Membrane Bio-Reactor + Nanofiltration/Reverse Osmosis) Combination Process for Landfill Leachate Treatment in China

Wu Yuhao[1]

1 College of Environmental Science and Engineering, Taiyuan University of Technology, Taiyuan, Shanxi, China
E-mail: 18058995738@163.com

Abstract. Landfill leachate involves high concentrations of organic compounds, ammonia nitrogen, heavy metals and other complex components. Hence, it is very difficult to treat using conventional biological processes alone. This research is a preliminary literature review on using MBR+NF/RO combination process to achieve purification of landfill leachate in China. MBR+NF/RO process is a combination of biological and physico-chemical technology, including MBR system, nanofiltration and reverse osmosis devices. The study found that, the benefit of a complementary combination of MBR and NF/RO process with respect to removal of ammonia nitrogen, trace organic and heavy metal ions appears quite intuitive. And this promising treatment has developed rapidly in China in last few decades. In this paper, representative project examples in China were selected for review. Mechanism and influencing factors of MBR+NF/RO process in China in the past decades will be reviewed and recommendations will be given for the future development of Chinese MBR+NF/RO process.

1. Introduction

In the past decades, the municipal solid waste (MSW) production has been increased rapidly throughout the whole world as an accompaniment of the increasingly wealthy lifestyles, continuing industrial and commercial growth in many countries [1]. Currently, landfill is the treatment which has been widely applied in most Chinese garbage disposal plant to treat MSW, compared to other treatments such as waste incineration, composting…, etc. In China, over 150 million tons of MSW are produced and keep a stable increase rate of 8%–10% every year. And about 80% MSW are disposed by local sanitary landfills [2, 3]. However, landfill leachate, a complex, hazardous wastewater, is formed during the biochemical, chemical and physical reaction process in the waste cells with the percolation of rainwater and the moisture contained in wastes and its composition fluctuates with the landfill age. The satisfactory purification of landfill leachate is hard to achieve by conventional processes alone, because it often contains a large quantity of organic matter, ammonia, heavy metals, and inorganic salts and varies of hazardous materials such as pharmaceuticals and personal care products (PPCPs) [3-6].

Therefore, a lot of physico-chemical and biological combination treatments process are studied and experimented by scientists domestic and abroad for effective leachate treatment [3, 6]. On the one hand, among all the biological treatment process, MBR has been expected to be a superior and effective way to treat landfill leachate due to its smaller footprint, higher effluent quality, increased biomass, process stability and low sludge production [4, 7-9]. Therefore, MBR process were widely applied in China for wastewater treatment with the average annual growth rate of nearly 100% and the capacity becoming larger and large in the past decades [10, 11]. On the other hand, a lot of physico-chemical processes for leachate treatment are intensively studied because of its resistance for toxic and hazardous materials. And NF/RO membrane process shows outstanding removal efficiency of colour, most organic contaminates and inorganic salt as a post-treatment of MBR for further desalination and purification [3]. In [12], Drewes et al. studied the efficacy of NF, RO and some conventional processes at removing organic carbon. It showed that TOC rejections by RO and NF were 94-96.4% and 91.3-94.5%, respectively, much higher than the conventional ones.

In order to meet those stricter environment quality standards (GB16889-2008), MBR and NF/RO process are combined together as a novel combination process for landfill leachate treatment [8]. A large amount of MBR+NF/RO combination process are put in use to treat landfill leachate both lab-scale and industrial scale in China. However, there is not such a detailed review on MBR+NF/RO combination process in China recently. In this paper, the mechanism and application situation of MBR+NF/RO combination process for landfill leachate treatment and purification in China were reviewed, and the perspective analysis of representative project examples was carried out as well to provide recommendations for future development.

2. Literature Review

2.1. Characterization and Mechanism

2.1.1. MBR Step

MBR is a combination of conventional activated sludge process and membrane separation technology. With efficient membrane filtration that holds small pore sizes of less than 0.1 mm inside, MBR can control over both solids retention times (SRTs) and hydraulic retention times (HRTs) separately and reduce sludge production, which makes MBR more advanced and efficient than other biological treatment [13, 14]. In [13], the performance of MBR and sequencing batch reactor (SBR) are compared in high-strength landfill leachate treatment. Result shows the removal efficiencies of MBR for BOD_5, TN and NH_3 surpass 95% with an improvement of SBR removal efficiencies ranging from 21 to 34%.

With the popularity of MBR in China, various kinds of MBR appear including anaerobic MBR, aerobic MBR and etc. However, the two-stage anoxic/oxic MBR process (AO/AO-MBR) is applied in most landfill leachate treatment project in China, due to its superior removal of nitrogen through nitrification and denitrification [15]. This MBR process contains of a two-stage anoxic and aeration tank followed by an ultrafiltration (UF) process to achieve further clarification. With the biodegradation reaction in the two-stage anoxic and aeration tank and the filtration of the UF, most of chemical oxygen demand (COD), ammonia (NH_4^+-N) and total nitrogen (TN) can be removed effectively. In [15, 16], Liu et al. observed a two-stage AO/AO-MBR for landfill leachate treatment for a long period (81 d, 113 d). The best performance showed that the removal efficiencies of COD, ammonia and TN reached 85.6%, 99.1%, and 77.6%, respectively. Thus, the two-stage AO/AO-MBR can serve as an effective secondary treatment in landfill leachate treatment and the primary step in the MBR+NF/RO combination process.

2.1.2. NF/RO Step

Despite of high removal rates of MBR process on COD, ammonia and TN, the effluent still can not meet the effluent discharge standards (GB16889-2008) as shown in **Table 1** [17]. And MBR also has its limitation in a lot of trace organic pollutants [18]. Hence, NF/RO process is adopted as the tertiary treatment for MBR effluent treatment due to its acceptably high removal efficiency on inorganic molecules, total solid and residual trace organic such as persistent organic pollutants (POPs), pharmaceutical active compounds (PhACs) and hormones [18, 19].

The key mechanisms of NF and RO membrane process are similar, mainly including steric hindrance, electrostatic interactions (repulsion), and hydrophobic interactions (adsorption) between the influent and the membrane [20]. The distinction between NF and RO is the level of rejection on varies of pollutants. The effects of NF are between UF and RO. However, on the one hand, the rejection of divalent ions and the flux of NF are relatively higher than those of RO [19]. On the other hand, RO can retain certain trace organic contaminants which NF, UF and MBR can not remove [18]. Therefore, these two membrane technologies are combined together and scientists always put NF before RO to prevent membrane fouling and maintain membrane lifespan in the real landfill leachate treatment projects in China.

2.2. Application of MBR+NF/RO Combination Process in China

2.2.1. Structure and Scale

In order to obtain a further purification for landfill leachate and meet the discharge standards of the National Municipal Waste Sanitary Landfill Pollution Control Standard, MBR and NF/RO processes are combined together in most landfill leachate treatment projects in China [3, 21-23]. In [3], Wang et al. studied the treatment of landfill leachate using MBR+NF/RO combination process with granular active carbon (GAC) as pre-treatment in Taizhou, Jiangsu. The system exhibited excellent removal of those main pollutants and low rate of membrane fouling. Zhang et al. reported the situation of a landfill leachate treatment plant in Baoding, Hebei in [21]. In this project, an anaerobic biological denitrification MBR+NF/RO was adopted for a better removal of ammonia and TN (exceeding 99%) and it is reported that the Methane produced in the anaerobic process can be used to generate electricity. In [22], a project for leachate treatment by using up-flow blanket filter (UBF) and MBR+NF/RO combination process in the south of China was studied. And the paper showed that the final effluent can meet the standards of "the reuse of urban recycling water-Water quality standard for industrial uses" (GB/T 19923—2005) and be reused to replenish circulating cooling water.

There are various structures and scales of this combination process in practical project and some of the representative project examples in China were selected to show in **Table 2** [3, 5, 17, 21-33]. Despite of all kinds of treatment projects, the core steps are still the MBR+NF/RO process which undertakes the main task of removing COD, trace organic, ammonia, TN, heavy metal, inorganic salts, color and so many other pollutants in landfill leachate [29]. The flow chart of the combination process is showed in **Fig. 1**. And more physico-chemical and biological process can be added as pre-treatment or post-treatment for water quality stability and membrane fouling remission.

2.2.2. Removal Efficiency

Standard for pollution control on the landfill site of municipal solid waste (GB16889-2008) strictly required the discharge standards for basic pollutants in landfill leachate and the specific standards can be found in **Table 3** . All the treatments and structures are designed for meeting the standard so that the effluent can be recycled and reused harmlessly. As the main body of the combination process, MBR can improve its removal efficiency of hydrophobic and biodegradable trace pollutants through the adsorption between the pollutants and the mixed liquor suspended solids (MLSS) which extends the retention time in the biological tank. Then the MBR effluent with some unqualified pollutants will flow into NF/RO system and most of them will be rejected by the membrane reaction [7]. From **Table 4**, the removal of the main pollutants in each step can be seen clearly. In general, the removal efficiency of each step is quite high and compatible. The content of various pollutants is weakened step by step, and finally meets the emission standard.

Some data about pollutants removal efficiency of several projects in various parts of China are collected in **Table 5** [3, 5, 17, 21-33]. It can be seen from the table that the leachate compositions are extremely unstable and will fluctuate with the actual situation. After comprehensive treatment of MBR+NF/RO combination process, the removal efficiency of COD and ammonia nitrogen is generally high, which can reach more than 99%. And the removal rate of TS is relatively high, even reaching 100% in [25]. This combination process can not only guarantee excellent removal effect, but also extend the efficient time through the self-optimization inside the system. Firstly, the MBR prior can reduce the membrane fouling problem of NF/RO system after precipitation and adsorption [7, 34]. Secondly, the application of NF/RO membrane system can increase the volume of water reclaimed and reduce the infiltration problem in some reusing area such as agricultural irrigation and etc. [35]

3. Future Prospects in MBR+NF/RO Combination Process in China

From various papers and data, it can be known that the MBR+NF/RO combination process has been widely applied throughout China. Nowadays, China's land resources are becoming increasingly tense, and regional development is extremely uneven. As the main technology of landfill leachate treatment in China, it has the advantage of smaller footprint, higher effluent quality and lower operating costs, which is very suitable for China's national conditions.

MBR technology has matured in engineering practice. The future research is mainly focused on the optimization of its processing effects, such as the following points. Firstly, membrane fouling is still the main problem affecting the treatment effect and membrane lifespan. Thus, how to eliminate membrane fouling more effectively should be studied further. Secondly, research on energy conservation and cost reduction should be encouraged. For example, methane produced during the anaerobic stage of MBR can be used to generate electricity to provide the pressure needed for RO step. Furthermore, the treatment of leachate concentrates formed during treatment should draw people's attention. The current treatment methods mainly include recharging advanced oxidation, etc., which are highly cost and harmful to the environment. Therefore, harmless treatment and recycling of leachate concentrate will be the focus of future research.

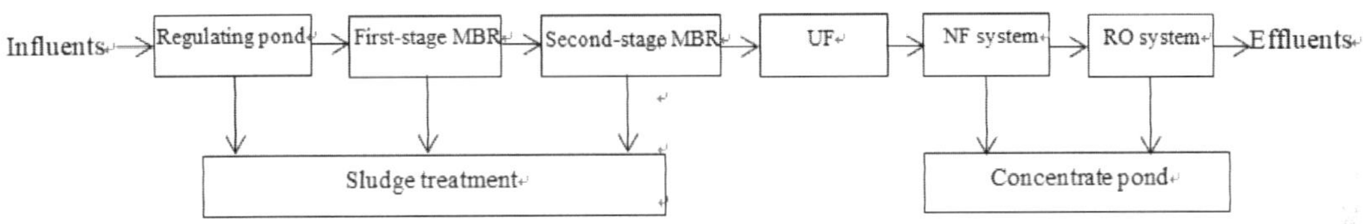

Fig 1. Flow chart of MBR+NF/RO combination process [29]

Table 1. Characteristics of the Raw Leachate and MBR Effluent. [17]

	Color	COD_{cr} (mg/L)	NH_4^+-N (mg/L)	Conductivity (mS/cm)
Raw Leachate	Black	4670-6700	820-960	14.9-15.3
MBR Effluent	Brown	568-850	0-2	0.78-11.4
GB16889-2008[a]	20 (10) times	100 (60)	25 (8)	-

[a] Standard for Pollution Control on the Landfill Site of Municipal Solid Waste. The data shown in the bracket is the first-class standard.

Table 2. Basic situation of landfill leachate treatment project examples in China.

Start year	Landfill site	Scale (m^3/d)	Process	Cost (yuan/m^3)	Ref.
-	Taizhou	1000	A/O-GAC-MBR+NF/RO	-	[3]
2005	Shanghai	950	A/O-MBR	-	[5]
2005	Tianjin	-	MBR+NF	-	[17]
2001	Baoding	150	A/O-MBR+NF/RO	23.24	[21]
2013	Suzhou	700	MBR+NF/RO	30	[22]
-	-	200-300	UBF[a]+MBR+NF/RO	-	[23]
2003	Chengdu	1300	MBR+RO	25	[24]
-	Qingdao	150	MBR+NF/RO	-	[25]
2009	Tengzhou	80	MBR+NF/RO	20-28	[26]
-	Changzhou	247	MBR+NF	19.55	[27]
-	Penglai	120	MBR+NF/RO	-	[28]
2001	Tianjin	200	MBR+NF/RO	36.5	[29]
2014	Yichang	200	TMBR[b]+NF/RO	29.5	[30]

2008	Chengdu	1000	MBR+RO/NF	-	[31]
2013	Chengdu	1000	MBR+NF	40-50	[32]
-	-	120	MBR+NF/RO	31.99	[33]

[a] Up-flow blanket filter.
[b] Tubular Membrane Bioreactor.

Table 3. Standard for pollution control on the landfill site of municipal solid waste, China, GB16889-2008.

Paramate	Emission limit value
Chromaticity (times)	40
COD_{cr} (mg/L)	100
BOD_5 (mg/L)	30
SS (mg/L)	30
NH_3-N (mg/L)	25
TN (mg/L)	40
TP (mg/L)	3
Hg (mg/L)	0.001
Cd (mg/L)	0.01
Cr (mg/L)	0.1

Table 4. Performance of different membrane process in treating effluent. [3]

Parameter	MBR	NF[a]	RO[b]	Std[c]	Std[d]
COD (mg/L)	579.7	276.25	12.39	100	60
NH_3-N (mg/L)	33	26.55	2.6	25	10
Chromaticity (times)	128	8	1	40	30
Conductivity (μS/cm)	4510	3740	134.22	-	-
Cd (mg/L)	0.047	ND	ND	0.01	-
Cr (mg/L)	0.17	ND	ND	0.1	-
Pb (mg/L)	0.12	ND	ND	0.1	-
Cl^- (mg/L)	1233.9	685.0	78.6	-	250
SO_4^{2-} (mg/L)	2334.6	166.6	55.3	-	250

ND: non-detectable.
[a] Operated under 1.4 MPa.
[b] Operated under 2.8 MPa.
[c] Standard for pollution control on the landfill site of municipal solid waste, China, GB8978-1996
[d] Reuse of recycling water for urban, water quality standard for industrial water consumpt-ion, China, GB/T 19923-2005

Table 5. Treatment effectiveness of landfill leachate treatment projects in China.

Landfill site	COD (mg/L)	BOD (mg/L)	pH	NH_3-N (mg/L)	SS (mg/L)	Removal (%)	Ref.
Taizhou	3134.88	450	7.85	434.76	-	99COD 99NH_3-N	[3]
Shanghai	15000-20000	7000-8000	7-8.5	2800-3300	1000-2000	94.2COD 95NH_3-N	[5]
Tianjin	4670-6700	-	-	820-960	-	99.1COD	[17]
Baoding	19854	6254	-	2894	-	99.7COD 99.7NH_3-N	[21]

-	4300	2000	3-4	700	5000	99.9COD 98.6NH_3-N	[23]
Chengdu	10000	3000	-	3000	800	99COD 99.2NH_3-N 96.3SS	[24]
Qingdao	12000	6000	5-8	12000	1000	99.5COD 99.2NH_3-N 100SS	[25]
Tengzhou	6300-8100	1200-2000	6.5-7.6	450-800	600-1000	99.6COD 99.4NH_3-N 99.9SS	[26]
Changzhou	5320-22000	-	-	1400-2300	-	99.5COD	[27]
Penglai	6000-9000	3000-5000	-	1200-1600	500-1500	99.6COD 99.8NH_3-N 98.5SS	[28]
Tianjin	6343-8216	-	6-9	1607-2147	-	99.7COD 99.9NH_3-N 96.3SS	[29]
Yichang	5001-18799	1500-8000	6.5-8.5	200-1000	1100-2000	99.6COD 95.8NH_3-N	[30]
Chengdu	2000-18000	1000-7000	-	1500-3000	-	99.8COD 99.8NH_3-N	[31]
Chengdu	3000-8000	1500-3000	-	2000-3500	1000-2000	98.2COD 99.6NH_3-N	[32]
-	11750	5482	-	924	745	99.2COD 96.8NH_3-N 96.6SS	[33]

4. Conclusion

The objective of this research was to have a primary study on the application of MBR+NF/RO combination process in China. A large number of domestic and foreign papers have been reviewed and analysed. Several of China's standards and specific projects have also been referenced.

MBR+NF/RO combination process has excellent adaptability, integrating the physico-chemical and biological treatment technology. Therefore, this treatment of all sizes can function well across China. In addition, MBR and NF/RO system can complement each other very well, reducing membrane fouling and achieving the discharge standards. And the efficiency of MBR+NF/RO process to remove COD and ammonia nitrogen is quite high and can exceed 99% in most engineering objects in China. Furthermore, the combination process in some special areas, such as the removal of antibiotics and hormones in landfill leachate, also has a good effect, which has important significance for the ecosystem. Finally, solving membrane fouling problems, energy conservation issues, and concentrate problem should be the direction of future research. Expectantly, this paper is comprehensive to determine the sufficient information and as a reference for further data collection to study landfill leachate treatment situation in China.

5. Reference

[1] Renou S, Givaudan JG, Poulain S, Dirassouyan F, Moulin P. Landfill leachate treatment: Review and opportunity. J Hazard Mater. 2008;150(3):468–93.

[2] Nie Y. Development and prospects of municipal solid waste (MSW) incineration in China. J Frontiers of Environmental Science & Engineering in China. 2008;2(1):1-7. Available from: https://doi.org/10.1007/s11783-008-0028-6

[3] Wang G, Fan Z, Wu D, Qin L, Zhang G, Gao C, et al. Anoxic/aerobic granular active carbon assisted MBR integrated with nanofiltration and reverse osmosis for advanced treatment of municipal landfill leachate. Desalination [Internet]. 2014;349:136–44. Available from: http://dx.doi.org/10.1016/j.desal.2014.06.030

[4] Chen S, Liu J. Landfill leachate treatment by MBR: Performance and molecular weight distribution of organic contaminant. Chinese Sci Bull. 2006;51(23):2831–8.

[5] Sui Q, Zhao W, Cao X, Lu S, Qiu Z, Gu X, et al. Pharmaceuticals and personal care products in the leachates from a typical landfill reservoir of municipal solid waste in Shanghai, China: Occurrence and removal by a full-scale membrane bioreactor. J Hazard Mater. 2017;323:99–108.

[6] Hasar H, Unsal SA, Ipek U, Karatas S, Cinar O, Yaman C, et al. Stripping/flocculation/membrane bioreactor/reverse osmosis treatment of municipal landfill leachate. J Hazard Mater. 2009;171(1–3):309–17.

[7] Alturki AA, Tadkaew N, McDonald JA, Khan SJ, Price WE, Nghiem LD. Combining MBR and NF/RO membrane filtration for the removal of trace organics in indirect potable water reuse applications. J Memb Sci [Internet]. 2010;365(1–2):206–15. Available from: http://dx.doi.org/10.1016/j.memsci.2010.09.008

[8] Ahmed FN, Lan CQ. Treatment of landfill leachate using membrane bioreactors: A review. Desalination [Internet]. 2012;287:41–54. Available from: http://dx.doi.org/10.1016/j.desal.2011.12.012

[9] Yamamoto K. Membrane bioreactor: An advanced wastewater treatment/reclamation technology and its function in excess-sludge minimization. Advances in Water and Wastewater Treatment Technology. 2001. 229-237 p.

[10] Wang Z, Wu Z, Mai S, Yang C, Wang X, An Y, et al. Research and applications of membrane bioreactors in China: Progress and prospect. Vol. 62, Separation and Purification Technology. 2008. p. 249–63.

[11] Zheng X, Zhou Y, Chen S, Zheng H, Zhou C. Survey of MBR market: Trends and perspectives in China. Desalination [Internet]. 2010;250(2):609–12. Available from: http://dx.doi.org/10.1016/j.desal.2009.09.034

[12] Drewes J, Reinhard M, Fox P. Comparing Microfiltration-Reverse Osmosis and Soil-Aquifer Treatment for Indirect Potable Reuse of Water. Vol. 37, Water research. 2003. 3612-3621 p.

[13] El-Fadel M, Hashisho J. A comparative examination of MBR and SBR performance for the treatment of high-strength landfill leachate. J Air Waste Manag Assoc. 2014;64(9):1073–84.

[14] Skouteris G, Hermosilla D, López P, Negro C, Blanco Á. Anaerobic membrane bioreactors for wastewater treatment: A review. Chem Eng J [Internet]. 2012;198–199:138–48. Available from: http://dx.doi.org/10.1016/j.cej.2012.05.070

[15] Liu J, Tian Z, Zhang P, Qiu G, Wu Y, Zhang H, et al. Influence of reflux ratio on two-stage anoxic/oxic with MBR for leachate treatment: Performance and microbial community structure. Bioresour Technol. 2018;256(January):69–76.

[16] Liu J, Zhang H, Zhang P, Wu Y, Gou X, Song Y, et al. Two-stage anoxic/oxic combined membrane bioreactor system for landfill leachate treatment: Pollutant removal performances and microbial community. Bioresour Technol [Internet]. 2017;243:738–46. Available from: http://dx.doi.org/10.1016/j.biortech.2017.07.002

[17] Li G, Wei W, Du Q. Applicability of nanofiltration for the advanced treatment of landfill leachate. J Appl Polym Sci. 2010;116(4):2343–7.

[18] Tay MF, Liu C, Cornelissen ER, Wu B, Chong TH. The feasibility of nanofiltration membrane bioreactor (NF-MBR)+reverse osmosis (RO) process for water reclamation: Comparison with ultrafiltration membrane bioreactor (UF-MBR)+RO process. Water Res [Internet]. 2018;129:180–9. Available from: https://doi.org/10.1016/j.watres.2017.11.013

[19] Mohammad AW, Teow YH, Ang WL, Chung YT, Oatley-Radcliffe DL, Hilal N. Nanofiltration membranes review: Recent advances and future prospects. Desalination [Internet]. 2015;356:226–54. Available from: http://dx.doi.org/10.1016/j.desal.2014.10.043

[20] Kim S, Chu KH, Al-Hamadani YAJ, Park CM, Jang M, Kim D-H, et al. Removal of contaminants of emerging concern by membranes in water and wastewater: A review. Chem Eng J [Internet]. 2017;335(September 2017):896–914. Available from: http://linkinghub.elsevier.com/retrieve/pii/S1385894717319563

[21] Zhang L, Jia Z, Liu J, Liang L. Application of anaerobic-biological denitrification-MBR-NF-RO process to the treatment of landfill leachate. Ind Water Treat [Internet]. 2015;35(7):96–8. (in Chinese) Available from: http://www.iwt.cn/EN/10.11894/1005-829x.2015.35(7).096

[22] Tian A, Yun-Bo W, Huang J, Cui X. Study on the Advanced Treatment of Waste Leachate by Using UBF-MBR-NF-RO technology. Environ Sci Technol. 2014;27(2):31–4. (in Chinese)

[23] Zhi-Hua L. Treatment of Leachate from MSW Incineration Power Plant with Pretreatment/Anaerobic Reactor/MBR/NF/RO Process. China Water & Wastewater. 2016;32(4):92–4. (in Chinese)

[24] Min HH, Yu DU, Liu SL, Yuan-Ning LI, Wang Q. MBR/RO Process for Treatment of Landfill Leachate. China Water & Wastewater. 2010;26(4):64–6. (in Chinese)

[25] Liu S, An W. Application of MBR-NF-RO in Landfill Leachate Treatment Project. Environmental Sci Manag. 2013;38(3):96–9. (in Chinese)

[26] Zhang X, Feng X, Xie Y. A project of MBR+two-membrane process for landfill leachate treatment. J Tianjin Univ Technol. 2010;26(8):31–5. (in Chinese)

[27] Huang J, Wang H, Jiao T, Yue-Zhong L, Zhu W. Research on Treatment Technology and Demonstration Projects of Landfill Leachate. Environ Sci Technol. 2008;21(10):35–8. (in Chinese)

[28] Tian-Tian FA, Sun YH, Yi LI, Tang SJ. Application of MBR in Landfill Leachate Treatment Project in Penglai City. China Water & Wastewater. 2012;28(8):72–5. (in Chinese)

[29] Wu Z, Wan Z, Polytechnic G. TREATMENT OF BEAN PRODUCT WASTEWATER BY PROCESSES OF UASB, TWO-STAGE CONTACT OXIDATION AND COAGULATION SEDIMENTATION. Technol Water Treat. 2014;40(11):129–35. (in Chinese)

[30] Wei T, Chen Y, Xiao Y, Peihong H. Application of TMBR and NF/RO Combined Technology in Landfill Leachate Treatment. Ind Saf Environ Prot. 2016;42(5):34–7. (in Chinese)

[31] Zhu Y, Liu BC, Yu DU, Guo-Qing L V. Design of Leachate Treatment Project in Chengdu olid Waste Landfill Site. China Water & Wastewater. 2010;26(18):73–5. (in Chinese)

[32] Guo-Qing L, Jiang H, Yue Z, Fu-Mei H, Sun Y. Design of Leachate Treatment Expansion Project in Chengdu Solid Waste Landfill Site. China Water & Wastewater. 2015 Sep;31(9):63–7. (in Chinese)

[33] Dong-Bing MA, Yue Z, Liu B, Tang MM, Yu DU, Jiang JM, et al. MBR+NF/RO Process for Treatment of Landfill Leachate. China Water & Wastewater. 2014;30(6):9–11. (in Chinese)

[34] Jacob M, Guigui C, Cabassud C, Darras H, Lavison G, Moulin L. Performances of RO and NF processes for wastewater reuse: Tertiary treatment after a conventional activated sludge or a membrane bioreactor. Desalination [Internet]. 2010 Jan 15;250(2):833–9. Available from: http://linkinghub.elsevier.com/retrieve/pii/S0011916409010510

[35] Gündoğdu M, Jarma YA, Kabay N, Pek TÖ, Yüksel M. Integration of MBR with NF/RO processes for industrial wastewater reclamation and water reuse-effect of membrane type on product water quality. J Water Process Eng [Internet]. 2018;(September 2017):0–1. Available from: http://linkinghub.elsevier.com/retrieve/pii/S2214714417305706

Low Cost and Long Durability Material for Water Treatment: Titania-Coated Cement

A. Mennad[1,2], B. Boutra[1]

1 Unité de Développement des Equipements Solaires, UDES/ Centre de Développement des Energies Renouvelables, CDER
RN no. 11 B.P. 386 Bou-Ismaïl, 42415 Tipaza, Algeria
2 Author to whom correspondence should be addressed
E-mail: mennad.abdelkader@udes.dz

Abstract. Primary studies have been carried out on titanium dioxide powder, and revealed the efficient photocatalytic property of this material. More techniques have been developed to deposit the titanium dioxide particles onto a substrate, thus preserving from any further filtration after a water treatment. White cement has been used, as both a binder and a substrate, to deposit onto it titanium dioxide powder and to carry out the photodegradation of the tartrazine in an aqueous solution. The experiment has been conducted into a photocatalytic reactor designed and registered as a prototype patent. The complete photodegradation of the tartrazine molecule occurs with the same way as observed with the titanium dioxide powder in the aqueous solution of tartrazine. The same catalytic material of the titania-coated cement has been submitted for a series of the photocatalytic degradation tests of the tartrazine molecule and remained still active and comparable to those observed during the first reaction tests.

1. Introduction

Initially used as a painting pigment additive, big resarch interests are expressed to the titanium dioxide due to the semiconductor character of the crystalline nanoparticles. The resulting studies confer many technological applications to the titanium dioxide, such as photocatalysis [1–6], electronics [7–9], medecine [10–14], optics [15–19], paper industry and environment depollution [2,20–25]. Besides its interesting photocalytic properties, the titanium dioxide powder is used as a support to improve the physical and consequently catalytic properties of supported catalysts in chemical reactions[3,12,26–28].

Many research studies have revealed the interesting photocatalytic properties of the titanium dioxide in application of water treatment and environment depollution. The water treatment consists to get rid of all impurities including the particles of the catalyst and the reuse of the purified water for a specific need. To get benefit of the photocatalytic properties of the titanium dioxide for the entire water purification without further treatment, it is necessary to deposit the particles of titanium dioxide onto a substrate. At this point, many techniques differing from either a chemical process such as CVD [29,30], ALD [31–34], SALD [35,36], or physical one like PVD [30,37–39] are used. These techniques require heavy equipments and need more investments.

Taking into account the limited budget and the materials, a low-cost technique to deposit titanium dioxide on cement can be achieved for a photocatalytic water treatment. This is the aim of this study.

2. Experimental

Raw materials:

IOP Conf. Series: Earth and Environmental Science **182** (2018) 012011 doi:10.1088/1755-1315/182/1/012011

White cement,

Deionized water

,Acetone, Analytical grade ACS

Titanium dioxide powder (BioChem Chemopharma product) , analytical reagent,

Tartrazine (68% purity)

Titanium dioxide deposition process:

The white cement is used as substrate onto which the aqueous solution of titanium dioxide is spread uniformally. An amount of white cement is kneaded with a mixture of the deionized water and Acetone (Acetone/H_2O = 1/10), and spread in a mold fitting the chamber dimensions of the photoreactor. A volume 100 ml of the colloidal solution of titanium dioxide (200 mg/l) is then spread onto the fresh surface of the white cement to cover the template dimensions 150 mm*100 mm*5 mm. The preparation is dried at ambient temperature over12 hrs. The plates coated with the catalyst are deposited at the bottom of the photoreactor to undergo the photocatalytic degradation of organic pollutants.
A soda lime glass of dimension 150 mm*100 mm*5 mm is used as a substrate to deposit 100 ml of the colloidal solution of TiO_2 (200 mg/l) onto the surface. Thus, a comparison of the adherence of the titanium dioxide particles with the two substrates, the cement and the soda lime glass, and eventually the catalytic behavior of the catalysts, can be observed.
Kinetic study of the photo degradation of tartrazine:

2 l volume of the aqueous solution of tartrazine (C_0= 20 mg/l) are treated by solar irradiation in continuous recycling mode by means of a peristaltic pump. The solution sample is analysed during the reaction by the absorbance measurements of the tartrazine molecule at the wave length λ= 427 nm by means of the spectrophotometer Shimadzu UV-1800. A preliminary spectrophotometric assay of the tartrazine helps to plot the absorbance versus the concentration and to calculate the equivalent concentration according to the Beer Lambert's equation1

$$A= \varepsilon l C \qquad (1)$$

Where:

A: absorbance, no units

ϵ: absortivity of the solution, $l.mg^{-1}.cm^{-1}$

l: length of the light path, cm

C: concentration of the tartrazine, mg/l

3. Results and Discussion

Prior to the kinetic study of the solar photodegradation of the tartrazine aqueous solution, a tartrazine spectrophotometric assay is reported as a plot of the absorbance versus the concentration in the figure 1.

Absorbance

IOP Conf. Series: Earth and Environmental Science **182** (2018) 012011 doi:10.1088/1755-1315/182/1/012011

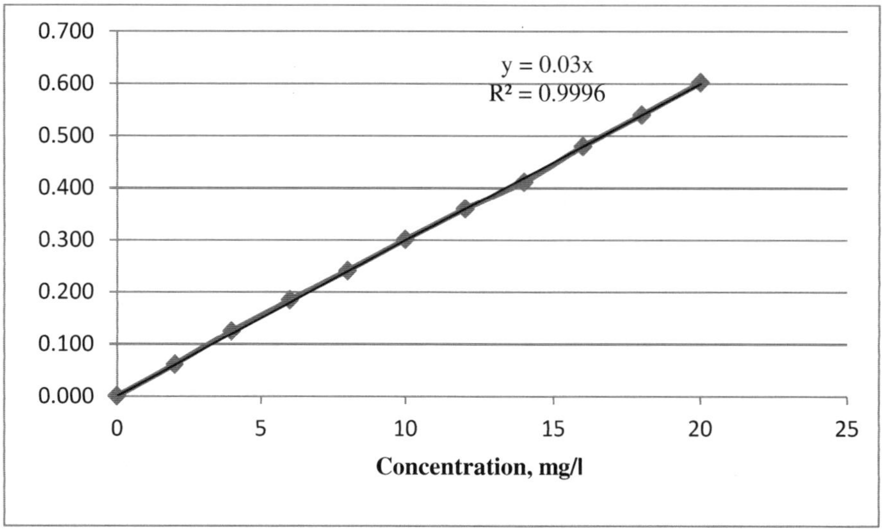

Figure 1. Tartrazine spectrophotmetric assay in aqueous solution at λ= 427 nm

The titania-coated cement substrate freshly prepared is deposited onto the bottom of the reactor and a volume 2 l of the aqueous solution of tartrazine (C_0 =20 mg/l) is added and exposed to the solar UV irradiation. A total degradation of the tartrazine molecule, as shown in the figure 2, is achieved after 210 min of solar UV irradiation. The same titania-coated cement substrate is exposed to the sun irradiation to dry and re-used for the second and third photocatalytic reaction tests of the tartrazine degradation. The results reported in the figure 2 show a stable activity of the re-used titania-coated cement substrate comparable to the freshly prepared one.
A soda lime glass substrate with the dimensions of 150 mm x100 mm x 5 mm is used to deposit 100 ml of the colloidal solution of the commercial BioChem Chemopharma titanium dioxide (200 mg/l, Acetone/H_2O=1/10) and submitted to the solar irradiation for the degradation of the volume 2 l of the tartrazine aqueous solution. The kinetic results related to the soda lime glass reported in the figure 3 are similar to those observed with the titania-coated cement (figure 2).
The repeated photocatalytic tests of the tartrazine degradation carried out on the titania-coated cement did not show any change of the catalytic and adherence aspect of the substrate.

Concentration

mg/l

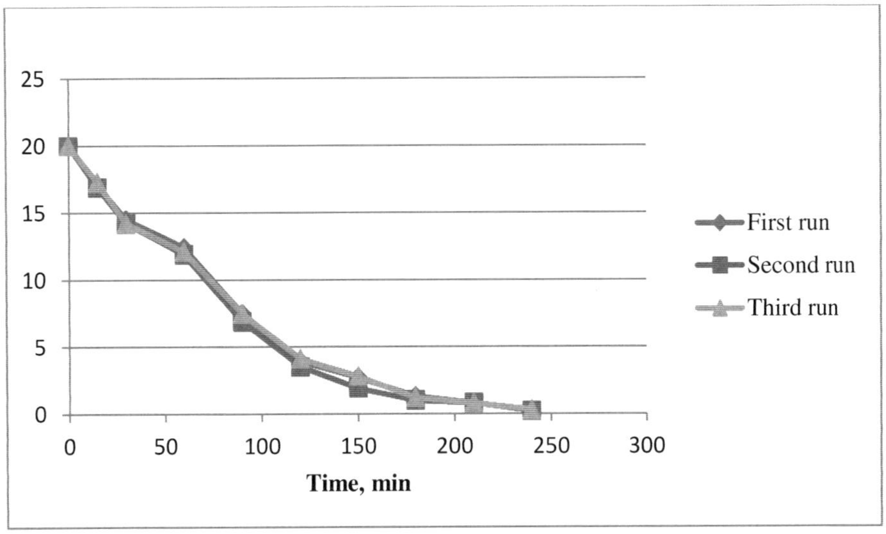

Figure 2. Solar photo degradation of the Tartrazine aqueous solution on titania-coated cement

Concerntration

mg/l

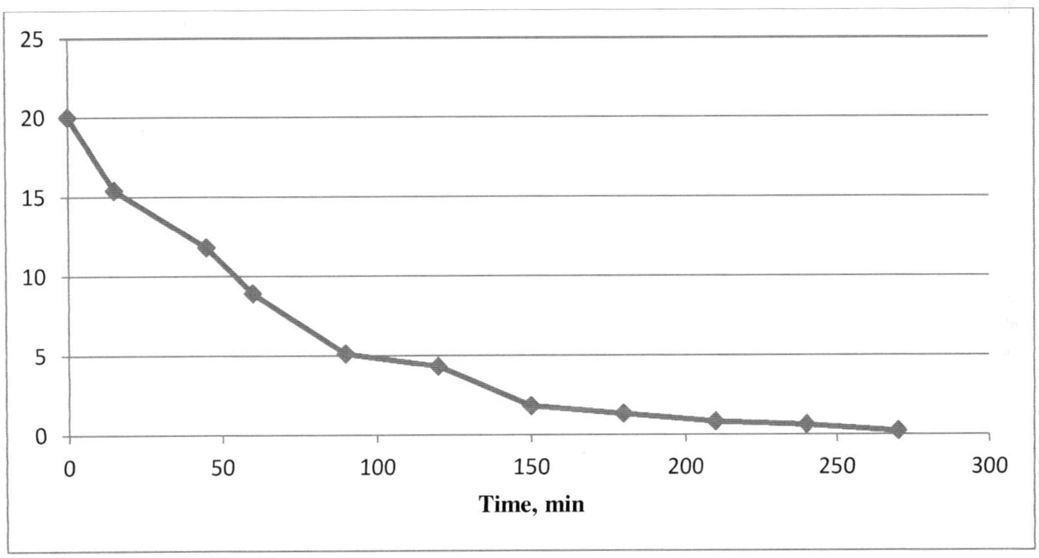

Figure 3. Solar photodegradation of Tartrazine aqueous solution on soda lime glass

4. Conclusions

The cement can be used, as both a binder and a surface substrate, to deposit the titanium dioxide catalyst. Many methods of deposition of the catalyst can be used such as spray, spread coating. The adherence of the titanium dioxide particles is stronger with the cemented substrate than with the soda lime glass.

5. References

[1] Cao Y, Xing Z, Hu M, Li Z, Wu X, Zhao T, Xiu Z, Yang S and Zhou W 2017 Mesoporous black N-TiO_{2-x} hollow spheres as efficient visible-light-driven photocatalysts *J. Catal.* 356 246–54

[2] Enesca A, Yamaguchi Y, Terashima C, Fujishima A, Nakata K and Duta A 2017 Enhanced UV–Vis photocatalytic performance of the CuInS2/TiO_2/SnO_2 hetero-structure for air decontamination *J. Catal.* 350 174–81

[3] Diak M, Klein M, Klimczuk T, Lisowski W, Remita H, Zaleska-Medynska A and Grabowska E 2017 Photoactivity of decahedral TiO_2 loaded with bimetallic nanoparticles: Degradation pathway of phenol-1-13C and hydroxyl radical formation *Appl. Catal. B Environ.* 200 56–71

[4] Friedmann D, Mendive C and Bahnemann D 2010 TiO_2 for water treatment: Parameters affecting the kinetics and mechanisms of photocatalysis *Appl. Catal. B Environ.* 99 398–406

[5] Turchi C S and Ollis D F 1990 Photocatalytic degradation of organic water contaminants: Mechanisms involving hydroxyl radical attack *J. Catal.* 122 178–92

[6] Fujishima A and Tryk D A 2009 *PHOTOCHEMICAL AND PHOTOELECTROCHEMICAL WATER SPLITTING*

[7] Velegraki T, Hapeshi E, Fatta-Kassinos D and Poulios I 2015 Solar-induced heterogeneous photocatalytic degradation of methyl-paraben *Appl. Catal. B Environ.* 178 2–11

[8] Tian-Hui Z, Ling-Yu P, Zhao S-L, Zheng X, Qian W and Chao K 2012 Application of TiO 2 with different structures in solar cells * *Chin. Phys. B* 21

[9] Zhang K, Wu C, Wang F, Miao Y, Liu K and Zhao J 2013 Effects of electrodes on resistance switching characteristics of TiO_2 for flexible memory *Optoelectron. Lett.* 9 263–5

[10] Bogdan J, Pławińska-Czarnak J and Zarzyńska J 2017 Nanoparticles of Titanium and Zinc Oxides as Novel Agents in Tumor Treatment: a Review *Nanoscale Res. Lett.* 12

[11] Lagopati N, Kitsiou P V., Kontos A I, Venieratos P, Kotsopoulou E, Kontos A G, Dionysiou D D, Pispas S, Tsilibary E C and Falaras P 2010 Photo-induced treatment of breast epithelial cancer cells using nanostructured titanium dioxide solution *J. Photochem. Photobiol. A Chem.* 214 215–23

[12] Xu F, Chen J, Kalytchuk S, Chu L, Shao Y, Kong D, Chu K H, Sit P H L and Teoh W Y 2017 Supported gold clusters as effective and reusable photocatalysts for the abatement of endocrine-disrupting chemicals under visible light *J. Catal.* 354 1–12

[13] Seo J, Chung H, Kim M, Lee J, Choi I and Cheon J 2007 Development of Water-Soluble Single-Crystalline TiO2 Nanoparticles for Photocatalytic Cancer-Cell Treatment *Small* 3 850–3

[14] Thevenot P, Cho J, Wavhal D, Timmons R B and Tang L 2008 Surface chemistry influences cancer killing effect of TiO2 nanoparticles. *Nanomedicine* 4 226–36

[15] Khan M I, Bhatti K A, Qindeel R, Althobaiti H S and Alonizan N 2017 Structural, electrical and optical properties of multilayer TiO2 thin films deposited by sol–gel spin coating *Results Phys.* 7 1437–9

[16] Mekprasart W, Khumtong T, Rattanarak J, Techitdheera W and Pecharapa W 2013 Effect of nitrogen doping on optical and photocatalytic properties of TiO_2 thin film prepared by spin coating process *Energy Procedia* 34 746–50

[17] Landmann M, Rauls E and Schmidt W G 2012 The electronic structure and optical response of rutile, anatase and brookite TiO_2 *J. Phys. Condens. Matter* 24 195503–10

[18] Shehap A M and Akil D S 2016 Structural and optical properties of TiO_2 nanoparticles/PVA for different composites thin films *Int. J. Nanoelectron. Mater.* 9 17–36

[19] Kitui M, Mwamburi M M, Gaitho F and Maghanga C M 2015 Optical Properties of TiO_2 Based Multilayer Thin Films: Application to Optical Filters *Int. J. Thin Fil. Sci. Tec. Int. J. Thin Film. Sci. Technol.* 4 17–21

[20] Garusinghe U M, Raghuwanshi V S, Batchelor W and Garnier G 2018 Water Resistant Cellulose-Titanium Dioxide Composites for Photocatalysis *Sci. Rep.* 8 1–13

[21] Quinones D H, Rey A, Allvarez P M, Beltran F J and Li Puma G 2015 Boron doped TiO_2 catalysts for photocatalytic ozonation of aqueous mixtures of common pesticides: Diuron, o-phenylphenol, MCPA and terbuthylazine *Appl. Catal. B Environ.* 178 74–81

[22] Wang X and Zhang, XiwangZhang T 2014 Recent Progress in TiO 2 -Mediated Solar Photocatalysis for Industrial Wastewater Treatment *Int. J. Photoenergy* 2014 1–12

[23] Thompson G, Swain J, Kay M and Forster C . 2001 The treatment of pulp and paper mill effluent: a review *Bioresour. Technol.* 77 275–86

[24] Paz Y 2010 Application of TiO_2 photocatalysis for air treatment: Patents' overview *Appl. Catal. B Environ.* 99 448–60
[25] Robert D and Malato S 2002 Solar photocatalysis: A clean process for water detoxification *Sci. Total Environ.* 291 85–97
[26] Zhang H, Kawamura S, Tamba M, Kojima T, Yoshiba M and Izumi Y 2017 Is water more reactive than H2in photocatalytic CO_2 conversion into fuels using semiconductor catalysts under high reaction pressures? *J. Catal.* 352 452–65
[27] Edwards J G, Davies J A, Boucher D L and Mennad A 1992 An Opinion on the Heterogeneous Photoreactions of N2 with H_2O *Angew. Chemie Int. Ed. English* 31 480–2
[28] Mennad A 1986 *Hydrocondensation du monoxyde de carbone en présence de catalyseurs au Fe et ou Ni supportés* (Université Claude Bernard Lyon 1)
[29] Li Puma G, Bono A, Krishnaiah D and Collin J G 2008 Preparation of titanium dioxide photocatalyst loaded onto activated carbon support using chemical vapor deposition: A review paper *J. Hazard. Mater.* 157 209–19
[30] Chen X and Mao S S 2007 Titanium Dioxide Nanomaterials: Synthesis, Properties, Modifications, and Applications *Chem. Rev.* 107 2891–959
[31] Aarik J, Aidla A, Uustare T, Ritala M and Leskela M 2000 Titanium isopropoxide as a precursor for atomic layer deposition: characterization of titanium dioxide growth process *Appl. Surf. Sci.* 161 385–95
[32] Ritala M, Leskela M and Rauhala E 1994 Atomic layer epitaxy growth of titanium dioxide thin films from titanium ethoxide *Chem. Mater.* 6 556–61
[33] Niskanen a, Arstila K, Leskela M and Ritala M 2007 Radical Enhanced Atomic Layer Deposition of Titanium Dioxide. Chemical Vapor Deposition *Chem. Vap. Depos.* 13 152–7
[34] Ritala M, Leskela M, Nykanen E, Soininen P and Niinisto L 1993 Growth of Titanium-Dioxide Thin-Films by Atomic Layer Epitaxy *Thin Solid Films* 225 288–95
[35] Poodt P, Cameron D C, Dickey E, George S M, Kuznetsov V, Parsons G N, Roozeboom F, Sundaram G and Vermeer A 2012 Spatial atomic layer deposition: A route towards further industrialization of atomic layer deposition *J. Vac. Sci. Technol. A Vacuum, Surfaces, Film.* 30 10802
[36] Levy D H, Nelson S F and Freeman D 2009 Oxide Electronics by Spatial Atomic Layer Deposition *J. Disp. Technol. Vol. 5, Issue 12, pp. 484-494* 5 484–94
[37] Mazzi A, Bazzanella N, Orlandi M, Edla R, Patel N, Fernandes R and Miotello A 2016 Physical vapor deposition of mixed-metal oxides based on Fe, Co and Ni as water oxidation catalysts *Mater. Sci. Semicond. Process.* 42 155–8
[38] Heinrichs J, Jarmar T, Wiklund U and Engqvist H 2008 Physical Vapour Deposition and Bioactivity of Crystalline Titanium Dioxide Thin Films *Engqvist Artif. Organs Trends Biomater. Artif. Organs* 22 104–10
[39] Ozkucur N, Wetzel C, Hollstein F, Richter E, Funk R H W and Monsees T K 2008 Physical vapor deposition of zirconium or titanium thin films on flexible polyurethane highly support adhesion and physiology of human endothelial cells *J. Biomed. Mater. Res. Part A* 89 57–67

Optimization of Heavy Metal Removal by Sulfate Reducing Bacteria in a High Concentration Zn-fed Fixed Bed Bioreactor Using Plackett Burman Design Experiments

Qiyuan Gu, Xinglan Cui, Xingyu Liu*, He Shang and Jiankang Wen

National Engineering Laboratory of Biohydrometallurgy, GRIMAT Engineering Institute Co., Ltd. (GRIMAT), 101407 Beijing, China
E-mail: guqiyuan@139.com, wellwoodliu@163.com

Abstract. This study evaluated the combined effect of six processing factors on Zn, Cd removal and sulfate reducing by sulfate reducing bacteria in a high concentration Zn-fed fixed bed bioreactor. Statistically valid Plackett Burman design of experiments was employed to carry out this study. The results obtained showed a high removal of Zn (99.63%), Cd (99.73%) at a maximum influent Zn concentration of 320 mg/L. Analysis of variance (ANOVA) of Zn and Cd removal revealed that the effect due to glycerol dosage and reflux ratio were highly significant (P value < 0.05). To establish the role of sulfate reducing bacteria (SRB) in the metal removal process, the characteristics of microbial community in fixed bed bioreactor was analyzed by High-throughput Sequencing during the steady operational process. The results obtained shows that the percentage of the dominant sulfate reducing bacteria in the sludge, such as *Desulfovibrio*, could reach 45.7%. Qualitative EDS analysis of the precipitate was performed. The results revealed that the precipitates as ZnS were confirmed by the EDS spectrum with strong peaks of zinc and sulfur.

1. Introduction

Heavy metal containing wastewater from mining and industrial processing can result in a serious problem toward the environmentand living things [1]. Compared to the organic pollutants, which can be biodegraded to harmless chemical species, heavy metals can be classified as toxic and non-biodegradation inorganic pollutants as they could accumulate in the food chain and get absobed by living organisms, including humans, which result in ion displacement and substitutionof essential ions form cellular sites and blocking of functional groups of important molecules and essentialnutrient transport systems [2]. Hense, it is very importance to remove heavy metals and sulfate from wastewater prior to its dicharge into the environment [3]. Conventional chemical precipitation treatment processes of heavy metal wastewater have been investigated all over the world including addition of lime to raise pH, aeration to oxidize ferrous iron and flocculating and so on. However, those processes produce metal-rich solid products, such as residual sludge, which need disposal in designated landfill sites and long-term management [4-6]. Therefore, an alternative approach of immobilization of heavy metals as sulfides using hydrogen sulfide generated by sulfate reducing bacteria (SRB) is now of considerable research interest [7].

Up to now, bacterial sulfate reduction has been studied to treat heavy metal containing wastewater in laboratory bioreactor including upflow anaerobic sludge blanket (UASB) and in pilot-scale tests [8-10]. This process utilizes SRB to reduce the sulfate to S^{2-}, HS^- and H_2S which could precipitate heacy metals in wastewater. SRB belong to a group of morphologically diverse and anaerobic

IOP Conf. Series: Earth and Environmental Science **182** (2018) 012012 doi:10.1088/1755-1315/182/1/012012

microorganisms and have an ability to use sulfate as the terminal electron acceptor and organic compounds as electron donors [11]. The process is briefly showed as follows:

$$\text{Electron donor} + SO_4^{2-} \rightarrow HS^- + HCO_3^- \quad (1)$$

$$Me^{2+} + HS^- \rightarrow MeS\downarrow + H^+ \quad (2)$$

$$HCO_3^- + H^+ \rightarrow H_2O + CO_2 \quad (3)$$

The SRB do their activity as oxidation-reduction reactions, in which the carbon is oxidized and the released electron from this reaction is used to reduce sulphate [12]. Heavy metal precipitation by SRB is influenced by a variety of parameters such as carbon resource concentration, pH, sulfate and heavy metal concentration [13]. Sevaral authors have reproted the effect of indivial parameter on heavy metals treatment. It is important to analyze and characterize the multi-parameters effect on the treatment performance. One way to study the combined effect is by using statistically designed experiments [14]. For example, individual main effect of parameters on a given response can be investigated by employing statisitcal experimental desigh technique, such as Plackett-Burman design [15].

In this study, different combinations of high and low levels of six operation parameters were chosen using the Plackett Burman design. ANOVA and Student's t test were used for statistical analysis of the results to assess the significance and effect of these parameters on Zn, Cd and sulfate reduction in the study.

2. Materials and Methods

2.1. SRB source and culture conditions

Excess sludge obtained from a municipal wastewater treatment plant was used as an inoculum. Element composition measured in the water phase of the sample used as the inoculum in this study comprised of 400 mg/L COD, 40 mg/L TN, 2mg/L TP, 5 mg/L Zn, 1mg/L Cd and 200 mg/L sulfate.

2.2. Experimental set-up

The schematic diagram of SRB system is presented in Figure1.

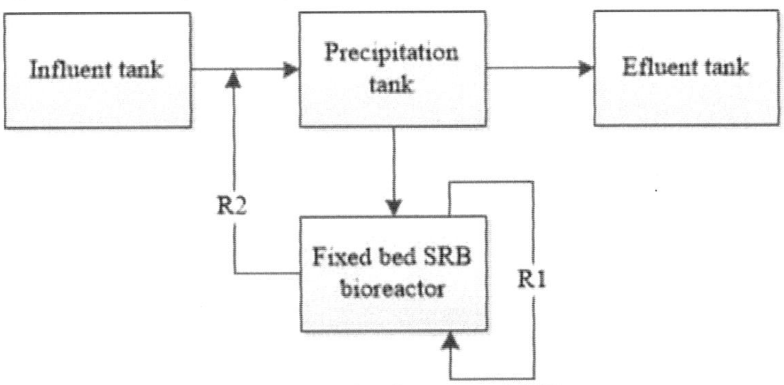

Figure 1. Schematic diagram of SRB system

The synthetic wastewater was pumped from influent tank to precipitation tank. Part of precipitation tank effluent was pumped to fixed bed SRB bioreactor. The fixed bed SRB bioreactor effluent which comprised of S^{2-}, HS^- and H_2S was refluxed to preciptation tank. The precipitation tank was equiped with a stirrer to promote the precipitaion reaction. The heavy metals precipitaion was carried on in the tank and R1 represent the reflux ratio between precipitation tank and SRB bioreactor. For satisfactory fluidization a reflux system was designed in SRB bioreactor and reflux ratio was represented as R2.

The SRB system were first inoculated with 10% (v/v) of the enrich SRB consortium and fed continuously with synthetic wastewater according to the experiment design plan.

2.3. Analytical methods

The concentrations of Zn and Cd were measured by an atomic absorption spectrophotometer after acidifying with concentrated nitric acid to prevent metal precipitation. Sulfate and pH were measured according to standard methods. The microstructure and qualitative analysis of precipitates were realized by scanning electron mircroscope(SEM) and energy dispersive spectromtry(EDS). The precipitates were obtained after filtering the sample from the bottom of precipitation tank througn 0.45 μm membrane filters. the filter paper was dried in an oven for 2 h at 105 °C and then thin coated by Au for SEM and EDS analysis.

2.4. Microbial community analysis
To study the microbial structure community in the SRB bioreactor, high-throughput sequencing technology was used. The total DNA of sample was extracted y PowerWater DNA Isolation Kit. After qualitative detection with agarose gel electrophoresis and determiantion by Nanodrop2000, the V4 hypervariable region of 16S rRNA gene was amplified with the primers of 515F (5'-GTGCCAGCMGCCGCGGTAA-3') and 909R (5'-CCCCGYCAATTCMTTTRA GT-3'). Gene sequencing was performed on an Illumina MiSeq platform by Sangon Corp method (Shanghai, China).

3. Results and Discussion

3.1. Heavy metal precipitation form SRB system
Table 1 presents the complete matrix of designed experiments and the results. The Minitab 17 was used to designed the experiments and analyze the data. The designed 12 experimental runs were carried out and the experimental data achieved was input to the software to calculate the predicted values. The results obtained reveal the the removal efficeincy of each of Zn, Cd and suflate varied depending upon the selected operation parameters. The results show a maximum removal for Zn(99.82%), followed by Cd(99.73%), and sulfate(40.00%). An overall removal efficiency of more than 80% for each of Zn and Cd was achieved. The results also show that experimental run 9 resulted in maximum sulfate reduction rate (40.00%), followed by that in in runs 4 and 10. The sulfate reduction efficiencies were the minimum in the experimental run 6.

Table 1. Plackett-Burman experimental design matrix.

Exp run	A Zn (mg/L)	B SO_4^{2-} (mg/L)	C Glycerol (mg/L)	D R1 (%)	E R2 (%)	F pH	Removal rate (%)		
							Zn	Cd	SO_4^{2-}
1	320	4000	250	200	200	5	93.13	91.33	11.00
2	220	2000	250	100	100	5	84.09	84.00	16.50
3	220	4000	250	100	100	6	87.50	85.60	20.00
4	320	4000	500	100	200	6	99.63	99.73	38.75
5	320	2000	250	100	200	6	99.28	94.67	20.00
6	320	4000	250	200	100	5	86.25	84.00	5.75
7	320	2000	500	100	100	5	94.56	94.13	34.00
8	220	2000	250	200	200	6	99.82	99.7	22.50
9	220	2000	500	200	200	5	98.64	96.60	40.00
10	220	4000	500	100	200	5	99.50	99.50	36.00
11	320	2000	500	200	100	6	98.94	98.6	33.00
12	220	4000	500	200	100	6	99.55	93.3	27.75

For a better interpretation and assesement of the significance of the selected operation parameters on Zn removal as well as on the Cd and sulfate reduction by SRB, the results obtained were analyzed statistically in terms of Student's test and analysis of variance(ANOVA).
The ANOVA of Zn, Cd and sulfate removal are presented in Table 2, 3 and 4 respectively. In these ANOVA tables, the high Fisher's F value and low probility P value demonstrate the precision of the

regression model in explaining the variatioins in the results. The value of Fisher's F and P illustrate whether the level means significantly different from each other or not.

Student's t test was used to illustrate the effect of the parameters on Zn, Cd and sulfate reduction (Table 5), which is often used to verify the sifnificance of the coefficients of the regression model parameter on the given response. The estimated coefficients of individual effect of these parameters and the associated T and P values were also showed inTable 5.

From Table 5, an increase in glycerol dosage showed positive effect with P values of 0.006, 0.007 and 0.000 on Zn, Cd and sulfate reduction, respectively. The parameter F (pH) showed a positive effect on Zn removal with a P value of 0.026. It was report that SRB could show better acitivty in basic pH range than in acidic Ph range. The influence due to parameters B(suflate concentration), D(R1) on sulfate removal were negative with P value of 0.049 and 0.058, whereas parameter E(R2) showed a positive effect with P value of 0.029. The increase in suflate concentration resulted in a negative effect on bacterial activity and also sulfate removal. Higher concentrations of sulfate reverse in the activity of the sulfate reducing bacteria, which could be due to inhibition fo metabolism and as a result inhibition of SRB growth at higher sulfate concentration. In addition, in the condition of high sulfate sulfate concentration the amount of disolved sulfide increase, therefore, the redox potential increases which causes inhibitory effect on SRB.

All these effects of parameters on Zn and Cd removal are better illustrated in the form of Pareto charts which is showed in Figure 2. The effect due to the individual parameter is represented by horizontal bars in the charts. And the those extending past the vertial line on the chart represent the significant ones (α=0.05). The results indicated that an increase in glycerol dosage and R2 showed positive effect on Zn and Cd removal. Glycerol dosage is a very important factor, because biological sulfate reduction involes two steps. In the first step, SRB oxidize glycerol by utilizing sulfate as an electron acceptor and generating hydrogen sulfide. In the seconde step, the biologically produced hydrogen sulfaide reacts with heavy metals to form metal sulfide precipitates.

Table 2. Analysis of variance of Zn removal

Source	DF	Seq SS	Adj MS	F	P
Main effects	6	353.719	58.953	8.58	0.016
A	1	0.601	0.601	0.09	0.779
B	1	7.965	7.965	1.16	0.331
C	1	138.353	138.353	20.13	0.006
D	1	11.520	11.520	1.68	0.252
E	1	127.368	127.368	18.53	0.008
F	1	67.912	67.912	9.88	0.026
Residual error	5	34.359	6.872		
Total	11	388.078			

Table 3. Analysis of variance of Cd removal

Source	DF	Seq SS	Adj MS	F	P
Main effects	6	358.715	59.786	7.56	0.021
A	1	1.182	1.182	0.15	0.715
B	1	16.882	16.882	2.13	0.204
C	1	150.993	150.993	19.09	0.007
D	1	2.901	2.901	0.37	0.571
E	1	146.301	146.301	18.50	0.008
F	1	40.456	40.456	5.11	0.073
Residual error	5	39.548	7.910		
Total	11	398.263			

Table 4. Analysis of variance of Sulfate removal

Source	DF	Seq SS	Adj MS	F	P
Main effects	6	1335.86	222.64	25.00	0.001
A	1	34.17	34.17	3.84	0.107
B	1	59.63	59.63	6.70	0.049
C	1	1078.26	1078.26	121.08	0.000
D	1	53.13	53.13	5.97	0.058
E	1	81.38	81.38	9.14	0.029
F	1	29.30	29.30	3.29	0.129
Residual error	5	44.53	8.91		
Total	11	1380.39			

Table 5. Significance of different parameters on contaminants removal in terms of Student's t-test

	Term	A	B	C	D	E	F
For Zn removal	Effect	0.448	-1.629	6.791	1.960	6.516	4.758
	Coef	0.224	-0.815	3.396	0.980	3.258	2.379
	T	0.30	-1.08	4.49	1.29	4.31	3.14
	P	0.779	0.331	0.006	0.252	0.008	0.026
For Cd removal	Effect	0.628	-2.372	7.094	0.983	6.983	3.672
	Coef	0.314	-1.186	3.547	0.492	3.492	1.836
	T	0.39	-1.46	4.37	0.61	4.30	2.26
	P	0.715	0.204	0.007	0.571	0.008	0.073
For sulfate reduction	Effect	-3.375	-4.458	18.958	-4.208	5.208	3.125
	Coef	-1.688	-2.229	9.479	-2.104	2.604	1.563
	T	-1.96	-2.59	11.00	-2.44	3.02	1.81
	P	0.107	0.049	0.000	0.058	0.029	0.129

Individual parameter effect, such as pH, heavy metal concentration and so on, on sulfate reduction by SRB is well reported in the literature. However, there is a very less understanding on the combined effect of more than one parameter on treatment performance. It can be observed from results that Zn and Cd precipitation by SRB through sulfate reduction. The obtained high Zn and Cd removal rates were due to its low solubility product with sulfide. It was reported that the order of removal of heacy metals as sulfide salts is attributed to their respective solubility product constant values. For example, the solubility product constant of Zn sulfide is 2×10^{-25}. The substantial precentage of Zn removal can be explained due to its low solubility product with sulfide. The other reason for high percentage of Zn removal is due to the feature of Zn for be bounded by extracellular polymeric substances (EPS) extracted from SRB. Charge density, attractive interaction and conformation types of the polymer are the parameters which determin the affinity of different metal ions.

The results further suggest that Zn at a high concentration of 320 mg/L did not inhibit the growth of SRB. At a high influent Zn concentration, the sulfate reducntion efficiency was slightly lower. An increase in Zn concentration showed negative effect with Stundent's t-test value of -1.96 on sulfate reduction. Possible expalnation may be the presence of bacteria cells facilitated metal precipitation and the high level of sulfide precipitates could act as a barrier between the cells and their essential growth nutrients. Therefore, the high influent heavy metal concentration could result in the drop of sulfate reduction. In addition, in others biological sulfate reduction process, the heavy metal containing wastewater was directly feeding into bioreactor, and in the reactor both sulfate reduction and metal precipitation occurs. In such process, high concentration of heavy metals will perform toxic effect on microbial activity, and result in the failure in treatment. In this studied novel process, by using water recycle (R2), the produced water soluble H_2S was pumped to precipitation tank and precipitated with metal to form metal sulfides, thus the heavy metals concentration and H_2S will maintain in low concentration. Thus only low concentration of heavy metals was fed into SRB bioreactor. Such process will decrease the toxicity of heavy metal to the microbial community.

The high tolerance of SRB in fixed bed bioractor was due to the biofilm struction. The metal resistance of SRB bilfilm varies with the species by developing a variety of specific resistance emchanism such as metal exclusion by permeablility barrier and reduction in metal sensitivity of biofilm. It was reported that the toxic effects of binary mixtures of heavy metal were obviously higher than the one of individual heavy metal. However, in the present study, 99~100% of Zn and Cd removal efficiencies were obtained during the experiments. The results might also be due to the use of an indigenous SRB consortium cultured from the municipal wastewater treament plant which accepted the industrial heavy metal containing wastewater. The advantage of using indigenous consortia may be that indigenous consortia contained muti-species which had adapted to a heavy metal polluted enviroment by developing a variety of resistence mechanisms.

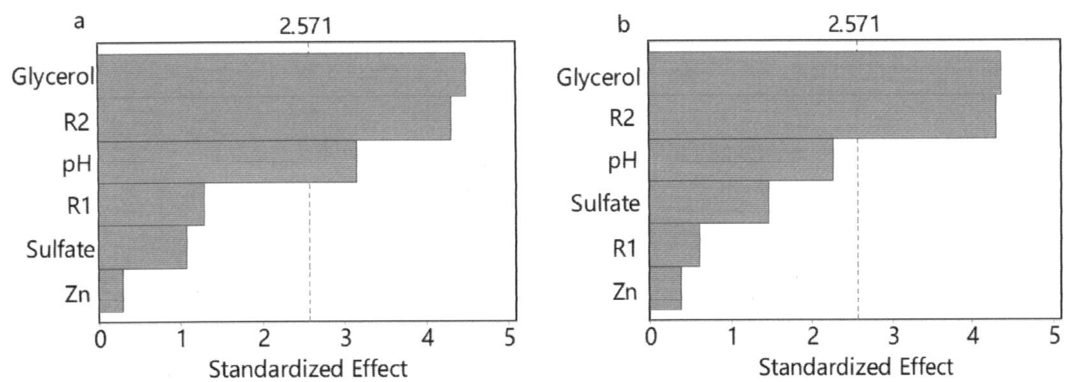

Figure 2. Pareto chart showing the effect of different parameters on Zn (a) and Cd (b) removal

3.2. Qualitative EDS analysis

In order to confirm the heavy metal removal mechanism by the SRB through sulfate reduction in this study, qualitative EDS analysis of the precipitates obtained from precipitation tank at the end of the experiments was performed.

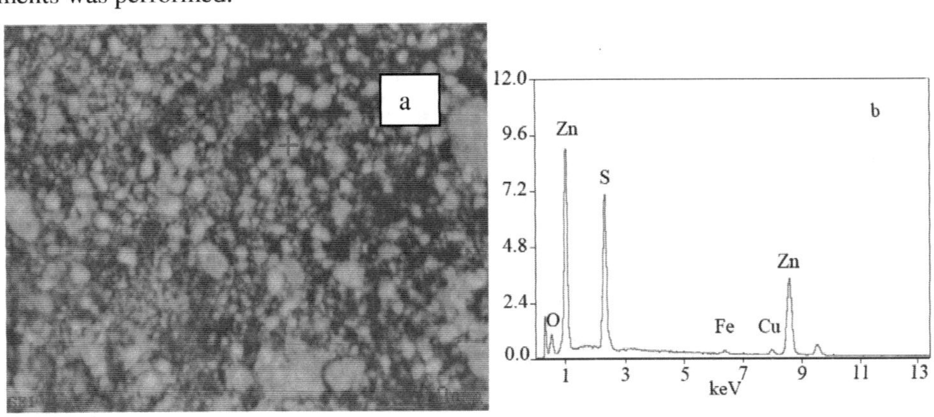

Figure 3. The SEM image (a) and EDS spectrum (b) of the precipitates

Metal precipitates are visible from Figure3(a). The precipitates as ZnS was confirmed by the EDS spectrum with strong peak of zinc and sulfur shown in Figure 3(b). The present of sulfur peak in the spectra in attributed to the metal sulfide precipitaion as a result of biological sulfate reduction reaction.

Sulfate reduction along with metal sulfide formation confirmed that sulfidogenesis is the predominant mechanism for heavy metal removal by SRB.

Metal sulfide have been attributed to the major precipitation of Zn by biological activity of SRB in the present study, which is confirmed by a simultaneous decrease of sulfate and Zn concentration in the effluent and EDS spectrum with strong peaks of zinc and sulfur. Among the peaks due to the different elements ,only the peak due to sulfide is significant indicating that metals were presented as sulfides in the precipitate. All other forms such as $M(OH)_x$, MCO_3 were not significant. The metal sulfide precipitates could be recoveried by smelting process.

The results indicated that Zn removal was possible only in the presence of sulfate and sulfide produced from sulfate reduction by SRB was responsible for Zn removal as ZnS. Although heavy metals may also have been removed through sorption to the biomassor by hydroxide and carbonate precipitation, sulfide precipitation is the dominant mechanism while other mechanisms play only a minor role for heavy metal removal in SRB system.

3.3. Analysis of SRB community

The microbial community structure in SRB bioreactor on genus level was showed in Figure 4. *Desulfovibrio* was the dominant microbe in the sample which was obtained from SRB bioreactor at experimental run 9. The detected potential SRB included *Desulfovibrio* (45.7%), *Sulfurospirillum* (1.6%), *Clostridium* (0.24%).

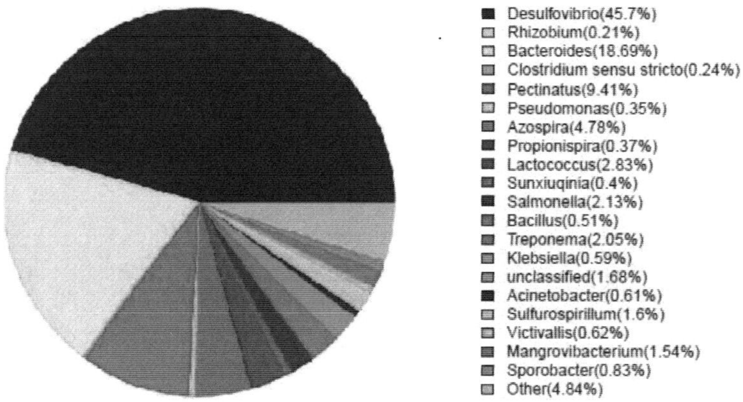

Figure 4. Microbial community structure on genus level in SRB bioreactor

It was reported that SRB was sensitive to acidity. When the pH drop from 3.25 to 3.0, the sulfate reduction efficiency decreased from 38.3% to 14.4%. Thus most of reported papers were operated at neutral pH. The decrease of pH value could result in the drop of sulfate reduction. In our study, *Desulfovibrio* became the dominant microbe under the condition of pH of 5.0~6.0. It was reported that the *Desulfovibrio* could survive at acid pH of 2.5. The works about isolating the acid- tolerant SRB will be carried on. The SRB technique operated under acid pH is very important, which will border the application area of this process.

4. Summary

This study demonstrated the precipitation of Zn and Cd at the condition of high influent Zn concentration of 320 mg/L. A very high removal of Zn and Cd was archieved at both low and high initial concentrations. Statistically valid Plackett Burman design of experiments was employed to carry out this study. Analysis of variance (ANOVA) of Zn and Cd removal revealed that the effect due to glycerol dosage and reflux ratio were highly significant (P value < 0.05). High-throughput sequencing technology was used to study the microbial community structure. The results indicated that SRB was the dominant microbe in SRB bioreactor. The precipitates as ZnS was confirmed by the EDS spectrum with strong peak of zinc and sulfur. Overall, this study proved very good potential of SRB in the anarobic biomass for metal precipitation with a high tolerance.

5. References

[1] Kiran M G, Pakshirajan K and Das G 2017 *J. Hazard. Mater.* 324 62
[2] Kieu H T Q, Muller E and Horn H 2011 *Water Res.* 45 3863
[3] Liu J, Zhou L, Dong F and Hudson-Edwards K A 2017 *Chemosphere* 168 529
[4] Yaneth V, Maria C E, Carmen M N, Ziv A and Fabio R 2016 *Chemosphere* 153 244
[5] Pakshirajan K and Kheria S 2012 *J. Environ. Manage* 101 118
[6] Guillermo P, Sebastien P, Keller J, Pablo L and Stefano F 2017 *Water Res.* 126 411
[7] Hong S, Cannon F S, Hou P, Byrne T and Cesar N D 2017 *Chemosphere* 184 429
[8] Pakshirajan K, Rene E and Swanminathan T 2009 *Int. J. Environ. Pollut* 36 266
[9] Bharati B and Ghosh P 2014 *J. Environ. Chem. Eng.* 2 1287
[10] Fan M, Lin Y, Huo H, Liu Y and Wei G 2016 *Water Res.* 96 198
[11] Choudhary R P and Sheoran A S 2011 *Bioresour. Technol.* 102 4319
[12] Jong T and Parry D L 2003 *Water Res.* 37 3379
[13] Pagnanelli F and Viggi C 2012 *J. Hazard. Mater.* 199 186
[14] Sahinkaya E and Yucesoy Z 2010 *Bioprocess Biosyst. Eng.* 33 989
[15] Kaksonen A, Franzmann P and Puhakka J 2004 *Biotechnol. Bioeng.* 86 332

Acknowledgments

This work was supported by the fund of General Research Institute for Nonferrous metals under grant numbers 53321 and 53348, and the public welfare fund of the Ministry of Environmental Protection of People's Republic of China under grant number 201509049, Program of International S & T Cooperation S2016G2135, the National Natural Science Foundation of China under grant numbers U1402234, 41573074.

Silver Nanoparticles Synthesised within the Silica Matrix in Hyperstoichiometrical of Mercury from Aqueous Solutions

A V Korobeinyk[1,2]* and V J Inglezakis[1]

1 Environmental Science & Technology Group (ESTg), Chemical Engineering Department, School of Engineering, Nazarbayev University, Astana, Kazakhstan
2 O. O. Chuiko Institute of Surface Chemistry of NAS of Ukraine, Kyiv, Ukraine
E-mail: alina.v.korobeinyk@gmail.com

Abstract. Mercury adsorption of silver containing silica-based nanocomposites was evaluated. Maximum adsorption capacity of 0.4 mmol g^{-1} was achieved at silver loading of 0.5 mmol g^{-1}. Nevertheless, if to calculate in respect to silver content the mercury adsorption capacity was generally elevated along with decreasing silver nanoparticle diameter. It has been demonstrated that silver particle diameters and loading should collectively be taken into consideration in designing the optimal mercury removal process. Further recommendations have been proposed with the aim of increasing the mercury removal efficiency using silver nanoparticles deposited on the surface of silica with lover silver loading, while achieving similar or even higher efficiencies due to observed hyperstoichiometry effect.

1. Introduction

Mercury and its compounds are persistent, highly bio-accumulative in to people and the environment [1] originates from natural and anthropogenic processes into environment in elemental, organic and inorganic forms [2]. They are classified as one of 13 priority hazardous substances (PHS) in accordance with the Water Framework Directive (WFD) and its daughter directives as for example the Environmental Quality Standards Directive (EQSD) which has been adopted recently [3, 4]. The maximum permissible concentrations (MPC) for mercury is 1.0 μg/L [5]. According to the current legislation, the EU will have had to stop using mercury in all the products and processes by 2020 [3]. Although this strategy will eventually lead to the desired result, i.e. a reduction of mercury pollution and achievement of new levels of MPC, but this unlikely occur soon. None of the existing commercially available technologies for mercury removal can reduce the level of mercury below the level of 50 ng/l [5]. Mercury exists in environment mainly in form of elemental mercury (Hg^0), inorganic (Hg^+, Hg^{2+}) and as organic species ($MeHg^+$, $MeHg^{2+}$, $EtHg^+$ and etc.) [6]. The constant redistribution of mercury species accrues in water as a function of temperature changes, evaporation and changes in the concentration of various inorganic and organic compounds, the speed and nature of water movement, the absorption of mercury by soil, plants and living organisms, etc. Water bed sediments acts as buffer media and depending on environmental conditions can accumulate or liberate mercury [7]. Due to the enormous toxicity of mercury and ability to accumulate in organisms, particularly in fish, it is a dangerous pollutant and one of the most studied. Nevertheless, there is a very incomplete understanding of the factors that control the bioconcentration and turnover of mercury in the living organisms.

Silver-mercury amalgamation is a process well known in a precious metal mining [8] and a formed amalgam alloy is the extremely inert chemically. The ability of silver nanoparticles to form such stabile compound might be useful in a mercury extraction from aqueous solution. In this work, silver nanoparticles of narrow size distribution were produced upon chemical redox transformation on a non-porous silica surface that was grafted with redox active silicon-hydride Si-H groups[9]; the only variable parameter was the metal salt concentration. The ligand free metal nanoparticles are converted to metallic nanoparticles and bound to silica surface via strong bonding interaction between the surface atoms of the nanoparticle and the Si-H groups of the SiO_2 matrix. Owing to its large specific surface area of 300–350 m^2/g and thermal stability, the fumed silica is an attractive material for utilization as a noble metal NP support [10]. Obtained silver NPs/SiO_2 composites were used in extraction of mercury from aqueous solutions.

2. Materials and Methods

2.1. Chemicals

$AgNO_3$ (>99.9%), SiO_2 (395 m^2/g), Triethoxysilane (TEOS, 95%), $HgCl_2$ (≥99.5%), acetic (glacial) and nitric (70.0%) acids, acetone and ethanol (all from Sigma-Aldrich) were used as received. Ultrapure water with resistivity of 18.3 MΩ·cm was obtained by reversed osmosis followed by ion-exchange and filtration (Millipore).

2.2. Silver metal nanoparticle synthesis

The silver NP/SiO_2 composites were prepared by *in-situ* redox reaction method. Silica surface was modified with TEOS in glacial acetic acid as described elsewhere[9] in order to create redox-active silicon-hydride functional groups. The common procedure for silver NPs/SiO_2 preparation was as follows: 1.0 g of TEOS-modified SiO_2 sample was immersed in 0.1 M aqueous solution of $AgNO_3$ with an appropriate volume to create metal NPs loading of 0.05, 0.2, 0.3, 0.4 and 0.5 mmol per gram of silica. The mixture was stirred in the dark for 1 h to the reaction completion. The mixture was then rinsed to remove unreacted salt, filtered and dried in a bench oven at 105 °C for 8 h.

2.3. Mercury adsorption

All solutions were prepared using ultra-pure water. Briefly, mercury solution was prepared by dilution of 1000 mg L^{-1} Hg stock solution ($HgCl_2$). The adsorption of Hg^{2+} on AgNP/SiO_2 nanocomposite was studied by batch technique. All batch adsorption experiments were performed by mixing a 1.0 g of nanocomposite sorbent into a $HgCl_2$ solution (8.5 mg/L, 100 mL) with stirring (120 r/min) at room temperature for 120 min. Hg solutions were adjusted to pH 6 by adding 0.1 M HNO_3. Then, the supernatant was separated from the mixture solution by filtration with 0.22 μm MF membrane. The concentration of Hg^{2+} ions remaining in solution was measured by mercury analyzer (RA-915+, Lumex). The amount of Hg^{2+} ions adsorbed by nanocomposite q_t (mg/g) is calculated using the following Eq. (1):

$$q_t = \frac{(C_0 - C_t) \cdot V}{m} \quad (1)$$

where C_0 and C_t are the initial and a time t liquid-phase Hg^{2+} concentrations (mg·L^{-1}), V is the volume of the solution (mL), and m is the mass of dry adsorbent used (g).

2.4. Characterization of materials

Wide-angle X-ray diffraction (XRD) was used to characterize the noble metal NP diffraction patterns. All measurements were collected over the 2θ range from 10° to 70° on a Rigaku SmartLab diffractometer equipped with a CuKα radiation source (0.1549 nm). A Nicolet 6700 FT-IR spectrometer (Thermo Scientific) with attenuated total reflection (ATR) attachment was used to collect IR spectra from silica samples after chemical modidfication. Transmission electron microscopy (TEM) analysis was carried out using a JEOL-JEM2100 instrument operated at 200 keV. All samples were ultrasonically dispersed in isopropyl alcohol at room temperature, and then a portion of the solution was placed onto a Cu grid for TEM measurements. Zeeman Mercury analyser (Lumex RA–915+ atomic absorption (AA) spectrometer coupled with a PYRO-915 pyrolysis attachment, Lumex Instruments) was used to analyse total mercury (THg) concentration [4].

3. Results and Discussion

Silica surface modified with TEOS contained Si-H surface groups which reduces silver ions when contacted with in aqueous solution of $AgNO_3$. According to the FTIR analysis (Figure 1, a) the chemical composition of the TEOS modified silica surface contain silicone-hydride groups (absorption band at 2240 cm^{-1} [11]. Iodometric titration results (not shown here) revealed the presence and total number of silicone hydride groups (0.83 mmol/g). Disappearance of the 2240 cm^{-1} band after reaction with Ag^+ (Figure 1, b) confirms consumption of those groups in reaction of silver reduction.

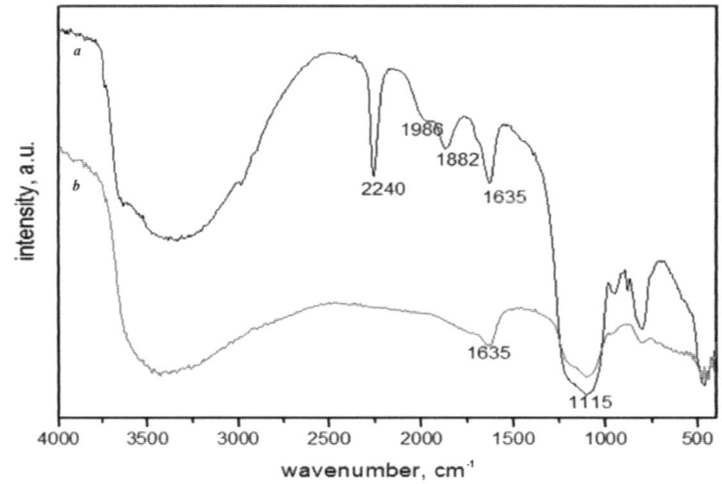

Figure 1. FTIR spectra of silica modified with triethoxysilane (a) and sample after interaction with silver ions (b).

After reduction of silver ions the original white silica sample gain a colour ranging from light yellow to dark ochre with increasing NP size and loading of samples from 0.05 mmol/g to 0.5 mmol/g. As shown in Figure 2b, virtually monodisperse nanoparticles were obtained at 0.05 mmol/g of Ag NPs loading. Their mean diameter was estimated by using the XRD spectra and the Debye–Scherrer formula to be 11 nm .

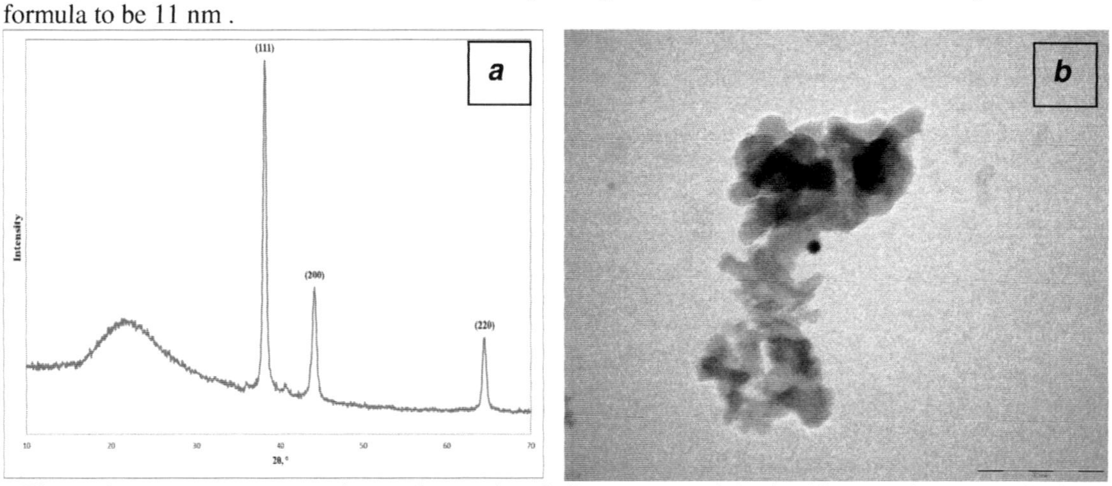

Figure 2. XRD spectra (a) and TEM image (b) of silver nanoparticles in silica matrix prepared with Ag NPs loading of 0.05 mmol/g.

XRD analysis of non-porous silica decorated with silver nanoparticles (Figure 1a) revealed a clear pattern of bands at 2θ values of 38.05°, 44.3° and 64.5° that correspond to (111), (200) and (220) planes of cubic Ag crystals, respectively. All the peaks in XRD pattern can be readily indexed to a structure of nanocrystalline silver (JCPDS, File No. 4-0783) [12]. The crystal structure of Ag NPs is well defined. A large halo at 22.5 2θ originates from the amorphous structure of silica. No peaks of other crystalline phases have been detected. All XRD spectra are similar to one presented in Figure 1, a.

According to the mercury adsorption experiments, nanocomposite samples reach equilibrium in 120 min (in whole range of silver loading). Therefore the time allowed for establishing equilibrium was 120 min with constant agitation. The values of Hg^{2+} adsorption uptake (Table 1) increased with the increase in silver content in composite, however if calculate mercury/silver ratio the composites with smallest AgNPs diameters – sample **1** (AgNPs/SiO_2_0.05) were found to have greater adsorption capacities than silica's with higher silver content under the experimental condition used. Samples of silica modified with TEOS and unmodified pristine silica sowed no affinity for mercury in tested conditions.

In order to prove applicability of silver containing silica in adsorption of mercury from "real" samples of water collected from heavily contaminated site – Lake-Reservoir Bulkyldak (Pavlodar city) the following experiment was designed: 1.0 g of AgNP/SiO_2_0.05 sample was introduced to the 100 mL of "real" mercury containing sample (14.84 ppb of total mercury analysed by RA-915+; pH adjusted to 6, by 0.1 M HNO_3) with stirring (120 r/min) at room temperature for 120 min. After 120 min of extraction the mercury concentration in supernatant was – 0.77 ppb (which is below of the MPC for mercury [5]), and extraction rate is 94.81%.

Table 1. Mercury adsorption results.

Sample	Ag content (mmol of Ag/g SiO_2)	Average AgNPs diameter (nm)	Hg uptake (mmol of Hg/ g of composite)	Hg/Ag ratio
1	0.05	11	0.1	2.0
2	0.2	24	0.2	1.0
3	0.3	37	0.25	0.83
4	0.4	63	0.3	0.75
5	0.5	90	0.4	0.8

4. Conclusions

This work demonstrated that the mercury adsorption characteristics of the silver decorated silica sample were influenced significantly by the size of silver nanoparticles. Amount of mercury adsorbed (mmol g^{-1}) in samples was deviated from stoichiometry, i.e. hyperstoichiometry was observed. The sample of silver containing composite demonstrated hyperstoichiometry effect was used in mercury extraction from "real" aqueous sample; reduction of mercury concentration below the MPC level was achieved. These findings demonstrate the need for further investigation on mercury adsorption mechanism in order to assess potential application in mercury removal from the aqueous solution.

5. References

[1] *Hazardous substances in Europe's fresh and marine waters: An overview*. 2011: Luxembourg. p. 66.

[2] Hsiao, H.-W., S.M. Ullrich, and T.W. Tanton, *Burdens of mercury in residents of Temirtau, Kazakhstan: I: Hair mercury concentrations and factors of elevated hair mercury levels.* Science of The Total Environment, 2011. 409(11): p. 2272-2280.

[3] US, E.P.A. *US EPA Method 7473: Mercury in Solids and Solutions by Thermal Decomposition, Amalgamation and Atomic Absorption Spectrophotometry. https://www.epa.gov/homeland-security-research/epa-method-7473-sw-846-mercury-solids-and-solutions-thermal-decomposition*. 2007b [cited 2017 December].

[4] Kolker, A., et al., *Mercury and trace element contents of Donbas coals and associated mine water in the vicinity of Donetsk, Ukraine.* International Journal of Coal Geology, 2009. 79(3): p. 83-91.

[5] D.W.Mazyck, A.M.Hagan, and H.Byrne, *Aqueous phase mercury removal: strategies for a secure future water supply. Critical National Need Idea White Paper.* , T.I.P. National Institute for Standards and Instrumentation, U.S. Department of Commerce 2009, Editor. 2009: Aug 2007, Washington, DC.

[6] Zhu, S., et al., *Speciation of mercury in water and fish samples by HPLC-ICP-MS after magnetic solid phase extraction.* Talanta, 2017. 171: p. 213-219.

[7] Covelli, S., et al., *Porewater Distribution and Benthic Flux Measurements of Mercury and Methylmercury in the Gulf of Trieste (Northern Adriatic Sea).* Estuarine, Coastal and Shelf Science, 1999. 48(4): p. 415-428.

[8] Leura Vicencio, A.K., L. Carrizales Yañes, and I. Razo Soto, *Mercury pollution assessment of mining wastes and soils from former silver amalgamation area in north-central Mexico.* 2017, 2017. 33(4): p. 15.

[9] Katok, K.V., et al., *Hyperstoichiometric Interaction Between Silver and Mercury at the Nanoscale.* Angewandte Chemie International Edition, 2012. 51(11): p. 2632-2635.

[10] Plumeré, N. and B. Speiser, *Redox-active silica nanoparticles: Part 2. Photochemical hydrosilylation on a hydride modified silica particle surface for the covalent immobilization of ferrocene.* Electrochimica Acta, 2007. 53(3): p. 1244-1251.

[11] Bellamy, L.J., *The Infra-red Spectra of Complex Molecules.* Vol. 2. 1980, London: Chapman and Hall. 299.

[12] Okitsu, K., et al., *Formation of noble metal particles by ultrasonic irradiation.* Ultrasonics Sonochemistry, 1996. 3(3): p. S249-S251.

Acknowledgments

The financial assistance of Nazarbayev University (ORAU projects SOE 2015 009 and 090118FD5319), Kazakhstan, European Commission, Horizon 2020 MSCA-RISE-2016 project 'NanoMed', and the Royal Academy of Engineering (UK), projects IAPP/1516/46 and IAPP/1516/3 research grants towards this research is hereby acknowledged.

Iodide Removal by Use of Ag-Modified Natural Zeolites

A V Korobeinyk[1,2]*, A R Satayeva[1], A N Chinakulova[1] and V J Inglezakis[1]

1 Environmental Science & Technology Group (ESTg), Chemical Engineering Department, School of Engineering, Nazarbayev University, Astana, Kazakhstan
2 O. O. Chuiko Institute of Surface Chemistry of NAS of Ukraine, Kyiv, Ukraine
E-mail: alina.v.korobeinyk@gmail.com

Abstract. In the present work Ukrainian clinoptilolite was modified with Ag and applied for the removal of iodide from aqueous solutions. The effect of three different modifications was studied , one resulting in an Ag^+ ion exchanged form, and two resulting in zeolites decorated with silver oxide and zero valent metallic nanoparticles. The results indicated the strong potential affinity of the Ag-modified zeolite materials towards iodide.

1. Introduction

Iodine is an essential element that's necessary for a normal thyroid function. However, high concentration of iodine in food and water calls a great concern of production of the congenital goiter. It should be pronounced that little attention has been paid to the role of iodine chemical form in its toxicity [1]. Authors [2] demonstrated that iodide has significant influence on the thyroid weight in male rats, although the iodide has low effect on the thyroid gland weight it changes thyroid hormones production level in both sexes, known as Wolff-Chaikoff effect [3]. WHO reports [4] an excessive iodine consumption related to numerous factors such as high level of table salt iodization or consuming iodine-rich food. Asian diet contain iodine-rich food [5], whereas in Europe drinking water with high level of iodine have been reported [6]. Combination of high iodine concentration in diet and drinking water with iodine supplementation leads to chronic iodide exposure resulting in iodism [7]. Iodine/iodide excess is a recognisable environmental risk factor for development of thyroid disorders and population of elderly, pregnant, foetuses and neonates are more susceptible to excess iodine induced diseases. Another concern with iodide/iodine is related to radioactive isotopes released from nuclear power plans. In the Fukushima Daiichi Nuclear Plant ancient (Japan, 2011), a large amount iodine radioisotope was released into the environment [8]. Therefore novel adsorbents towards to iodine/iodide removal should be developed in order to reduce their concentration in drinking water.

The use of zeolites (aluminosilicate porous natural materials) as adsorbents has attracted considerable attention due to their structural and chemical properties. Clinoptilolites have been found in many geological areas around the world and widely used in various applications [9]. A natural Ukrainian zeolite-Sokyrnit (NZU) is the mineral of volcano origin deposit in Transcarpathian region of Ukraine [10], which chemical composition is $(Na, K)_4\ CaAl_6Si_{30}O_{72}{\cdot}24H_2O$ [11]. Trasnscarpathian zeolites have been successfully applied in the environmental protection chemistry and industry [12]. Their non-toxic and non-swelling properties, thermal stability and potential of regeneration as well as high rate of ion exchange makes them excellent candidates in sorption processes. Clinoptilolites have been widely used in adsorption of cationic species due to negatively charged surface and presence of exchangeable cations (Na^+, K^+, Ca^{2+} and Mg^{2+}) [13], however they show a poor affinity towards anions [14]. Surfactants and noble metal species can be used to enhance clinoptilolite affinity towards to iodide [15, 16]. I^- anions are readily react with silver oxide (Ag_2O) to form the water-insoluble silver iodide [17]. Therefore, the aim of this study is to tailor the surface properties of NZU for iodide removal. The modification procedures were based on the use of $AgNO_3$ impregnation followed by thermal treatment and redox reaction with sodium borohydride in case of AgNPs formation. Preliminary adsorption study was carried out for water deiodification with modified and pristine NZU.

2. Materials and Methods

2.1. Materials

To study iodide adsorption NZU of particle size of 2-5 mm was supplied by the TM "Transcarpathian zeolitic factory" (Khust district, Ukraine). Chemical reagents such as NaCl (≥99.0%), KI (≥99.0%), $AgNO_3$ (>99.9%) and $NaBH_4$ (99.0%) (all from Sigma-Aldrich) were used as received. Ultrapure water with resistivity of 18.3 MΩ·cm was obtained by reversed osmosis followed by ion-exchange and filtration (Millipore).

2.2. Ion exchanged zeolites

The zeolite batch was firstly dried at 150 °C under for 24 h to remove the material humidity. The powder sample was sieved in a vibratory sieve shaker AS200 (Retsch), samples with particle diameters ranging from 0.8 to 1.6 mm were collected and used for ion exchange with NaCl. Homo-ionic Na-NZU was obtained via impregnation zeolite with 1 M NaCl for at 60°C in regards to the procedure described elsewhere [18]. Obtained homo-ionic clinoptilolite (NZU_Na) was washed with DI water in order to remove the occluded NaCl, filtered, dried at 60°C and stored in desiccator for the further usage. Ag^+ substituted zeolite was prepared from NZU_Na via impregnation with 0.04 M $AgNO_3$ in a light-shielded conditions [19]. In typical procedure 100 g of NZU_Na was impregnated with 250 ml of 0.04 M $AgNO_3$ at ambient temperature for 24 h, and then obtained zeolite (NZU_Ag^+) was filtered and dried in the bench oven for 8h at 60 °C.

2.3. Silver oxide-impregnated zeolites

Deposition of Ag_2O over the NZU_Na surface was performed as described in [20]: 3 g of NZU_Na was dispersed into the $AgNO_3$ solution (0.1 M, 0.2 L) under continuous agitation for 48 h. Sample then was filtered and dried in the bench oven for 24h at 60 °C and labeled as NZU_Ag_2O.

2.4. Silver metal nanoparticles-impregnated zeolites

The three-steps procedure for Ag nanoparticle synthesis on the surface of NZU was derived from [20]. Prior to the modification NZU_Na was calcinated ta 300°C for 3h. First two stages of synthesis were performed via subsequent saturation with 10 mM $AgNO_3$, in light-shield condition at ambient temperature for 8h, followed by drying at 180°C for 6h. To reduce the Ag_2O formed of the NZU surface the borohydride reduction method was adapted from [21]; for each batch of Ag_2O impregnated NZU (10 g) 20.0 mL of 50 mM $NaBH_4$ was used. The mixture was stirred in the dark for 8 h to the reaction completion; afterwards sample was rinsed to remove unreacted salt until pH neutral, filtered and dried in a bench oven at 90 °C for 8 h and stirred in desiccator prior to the use.

2.5. Iodine adsorption

All solutions were prepared using ultra-pure water. Briefly, iodide solution was prepared by dilution of 1000 mg L^{-1} I^- stock solution (KI). Batch experiments of iodide adsorption were carried out by immersion of zeolite samples equivalent to 0.1 g dry weight with 10mL of potassium iodide (100 ppm) solution. This was performed in PP centrifuge tubes fitted with caps, at 25 °C for 24 h [22]. Then, the supernatant solution was separated from the mixture solution by filtration with 0.22 μm MF membrane. The concentration of I^- ions remaining in solution was determined by ion chromatography (Model 930 Compact IC Flex, Metrohm, Switzerland). The amount of I^- ions adsorbed by zeolite q_t (mg/g) is calculated using the following Eq. (1):

$$q_t = \frac{(C_0 - C_t) \cdot V}{m} \qquad (1)$$

where C_0 and C_t are the initial and a time t liquid-phase I^- concentrations ($mg \cdot L^{-1}$), V is the volume of the solution (mL), and m is the mass of dry sample used (g).

2.6. Characterization of materials
The specific surface area properties of the natural and modified NZU were determined using an Autosorb 1C (Quantachrome Instruments). Chemical composition of pristine and modified samples was determined using X-ray fluorescence (XRF) spectroscopy using a PANalytical Axios mAX instrument and samples were prepared in an automatic fusion system (XrFuse 6, XRF Scientific, Brussels, Belgium).The adsorption isotherms of N_2 were determined at the temperature of the liquid nitrogen. Prior to its analysis, the samples were degasified for 24 h at 120 °C. Wide-angle X-ray diffraction (XRD) was used to characterize the zeolite diffraction patterns. All measurements were collected over the 2θ range from 10° to 90° on a Rigaku SmartLab diffractometer equipped with a CuKα radiation source (0.1549 nm). Ion chromatograph (930 Compact IC Flex) was used to analyse iodine concentration.

3. Results and Discussion
The nitrogen adsorption-desorption isotherm (not shown here) are classified as isotherms of IV Type and the BET surface area (Table 1) of raw NZU is small (12.6 m^2/g). Based on the nitrogen adsorption-desorption study (and confirmed with XRD) it can be stated that the clinoptilolite skeleton does not change with modification procedure, but average pore size changes as result of cation exchange (decrease) and sodium borohydride treatment (increased).

Table 1. Porosimetry analysis of natural and modified zeolites.

Samples	Porosimetry data		
	S_{sp} (m^2/g)	V (cc/g)	D_P (nm)
NZU	12.6	0.027	5.8
NZU_Na^+	15.0	0.032	2.98
NZU_Ag^+	14.9	0.076	3.15
NZU_Ag_2O	8.7	0.061	5.4
NZU_AgNPs	11.2	0.071	11.0

Mineralogical composition of each sample was determined by XRD analysis (Figure 1). Natural zeolite is primary comprised of clinoptilolite, while XRD spectra of modified samples contained peaks of silver corresponding to the formation of Ag-forms of clinoptilolite and heulandite (Table 2).

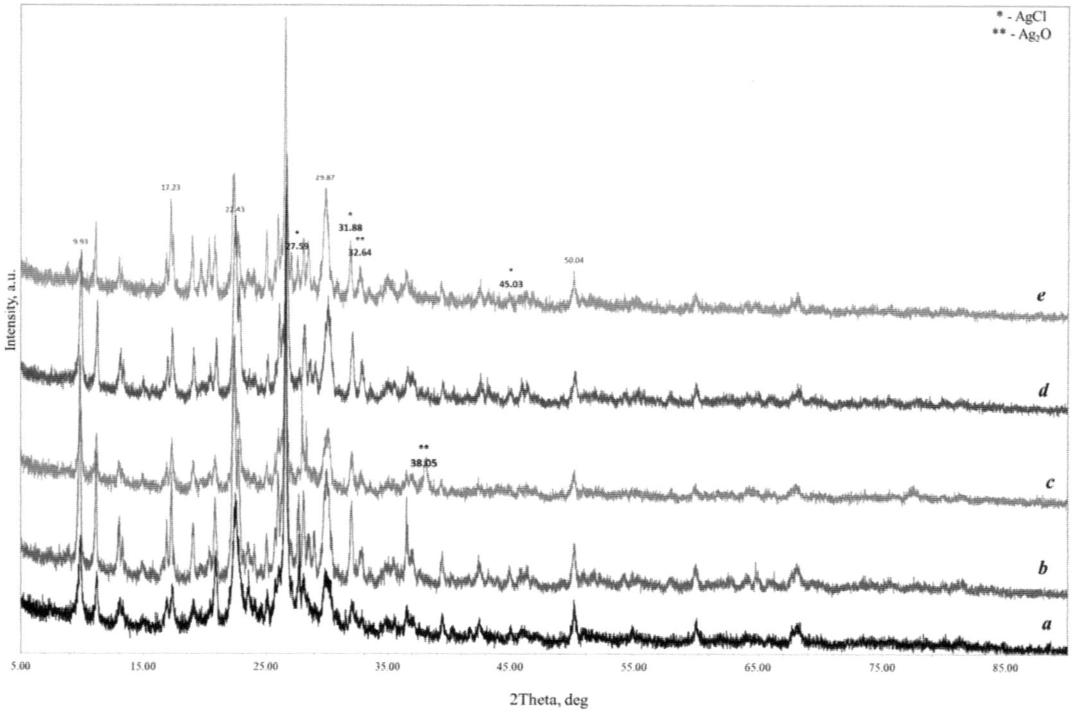

Figure 1. XRD patterns of NZU (a), NZU_Na (b), NZU_AgNPs(c), NZU_Ag^+(d) and NZU_Ag_2O (e).

XRD spectra of NZU-Ag_2O (Figure 1, e) shows bands of silver at 27.59, 31.88 and 45.03 2θ° (ref. file 01-075-9577, Chlorargyrite). The peak at 38.05 2θ in spectra of NZU_AgNPs (Figure 1, c) is characteristic of Ag metal (JCPDS, File No. 4-0783), which confirms the formation of Ag nanoparticles on the surface of zeolite.

Table 2. XRD analysis of zeolites.

Samples	Crystalline structure	
	Compound name	JCPDS, File No.
NZU	Clinoptilolite	01-079-1461
NZU_Na^+	Clinoptilolite-Na	01-079-1461
NZU_Ag^+	Clinoptilolite-Ag	01-081-8531
NZU_Ag_2O	Clinoptilolite-Ag	01-081-8531
NZU_AgNPs	Heulandite (Ag)	01-084-0419

According to the iodine adsorption experiments, zeolite samples which are non-loaded with silver had statistically insignificant adsorption values (Table 3, samples NZU and NZU_Na^+). The values of I^- adsorption extraction rate (Table 3) increased with the increase in silver content in composite, however this increase is non-monotonic and an influence of silver loading, particle size and chemical composition on the iodine extraction should be investigated further.

Table 3. Iodine adsorption results.

Sample	Ag content* (wt%)	Iodine extraction rate (%)
NZU	0.0	0.53
NZU_Na^+	0.0	0.72
NZU_Ag^+	0.91	67.0 (±0.3)
NZU_Ag_2O	7.9	100.0 (±0.5)

NZU_AgNPs	2.2	79.0 (±0.4)

* - determined by X-ray fluorescence (XRF).

4. Conclusions

This study demonstrates facile and low-tech method of natural Ukrainian clinoptilolite surface modification with $AgNO_3$. A subsequent physical-chemical treatment of Ag^+ impregnated zeolites leads to formation silver species (Ag_2O and AgNPs). Zeolite modification was confirmed by XRD analyses, without observing significant changes in zeolite skeleton; nevertheless, reducing of silver species on the zeolite surface with sodium borohydride leads to doubling in pore diameters. The adsorption properties of pristine and modified zeolites in aqueous solution towards iodide anion were studied and sample NZU_Ag_2O gave the best results. Ag-modified forms of clinoptilolite are promising materials for iodine removal from aqueous solution. Implementation of such synthetic approach paves the way for further research on the mechanism of anionic extraction.

5. References

[1] Medeiros - Neto, G.A., et al., *Congenital goitre and hypothyroidism with impaired iodide organification and high thyroid peroxidase concentration.* Clinical Endocrinology, 1979. 11(2): p. 123-139.

[2] Sherer, T.T., K.D. Thrall, and R.J. Bull, *Comparison of toxicity induced by iodine and iodide in male and female rats.* Journal of Toxicology and Environmental Health, 1991. 32(1): p. 89-101.

[3] Eng, P.H.K., et al., *Escape from the Acute Wolff-Chaikoff Effect Is Associated with a Decrease in Thyroid Sodium/Iodide Symporter Messenger Ribonucleic Acid and Protein**. Endocrinology, 1999. 140(8): p. 3404-3410.

[4] Zimmermann, M., *Iodine Deficiency and Excess in Children: Worldwide Status in 2013.* Endocrine Practice, 2013. 19(5): p. 839-846.

[5] Teas, J., et al., *Variability of Iodine Content in Common Commercially Available Edible Seaweeds.* Thyroid, 2004. 14(10): p. 836-841.

[6] Andersen, S., S.B. Petersen, and P. Laurberg, *Iodine in drinking water in Denmark is bound in humic substances.* European Journal of Endocrinology, 2002. 147(5): p. 663-670.

[7] Rivero, J.D., C. Cochran, and F.S. Celi, *Acute ioderma: An uncommon serious adverse event associated with saturated solution of potassium iodide treatment.* AACE Clinical Case Reports, 2018. 4(2): p. e127-e130.

[8] Xu, S., et al., *Iodine Isotopes in Precipitation: Temporal Responses to 129I Emissions from the Fukushima Nuclear Accident.* Environmental Science & Technology, 2013. 47(19): p. 10851-10859.

[9] Stocker, K., et al., *Characterization and Utilization of Natural Zeolites in Technical Applications.* BHM Berg- und Hüttenmännische Monatshefte, 2017. 162(4): p. 142-147.

[10] Moklyachuk, L., et al., *The use of natural zeolites impregnated with zinc and copper ammoniates as carriers of micronutrients in growing vegetables.* Emirates Journal of Food and Agriculture, 2013. 25(12): p. 980-985.

[11] Tarasevich, Y.I., et al., *Physical and Chemical Properties of the Transcarpathian zeolite and its application as a filter material in water treatment.* Chem. Tech. Water, 1979. 1(1): p. 66-69.

[12] Vasylechko, V.O., et al., *Adsorption of Copper on Transcarpathian Clinoptilolite.* Adsorption Science & Technology, 1999. 17(2): p. 125-134.

[13] Delkash, M., B. Ebrazi Bakhshayesh, and H. Kazemian, *Using zeolitic adsorbents to cleanup special wastewater streams: A review.* Microporous and Mesoporous Materials, 2015. 214: p. 224-241.

[14] Loganathan, P., et al., *Defluoridation of drinking water using adsorption processes.* Journal of Hazardous Materials, 2013. 248-249: p. 1-19.

[15] Chebbi, M., et al., *Influence of structural, textural and chemical parameters of silver zeolites on the retention of methyl iodide.* Microporous and Mesoporous Materials, 2017. 244: p. 137-150.

[16] Yu, F., et al., *Enhanced removal of iodide from aqueous solution by ozonation and subsequent adsorption on Ag-Ag2O modified on Carbon Spheres.* Applied Surface Science, 2018. 427: p. 753-762.
[17] Wiberg, E., N. Wiberg, and A.F. Holleman, *Holleman-Wiberg's Inorganic Chemistry*. 1st ed. 2001, New York: Academic Press. 1884.
[18] Lihareva, N., et al., *Ag+ sorption on natural and Na-exchanged clinoptilolite from Eastern Rhodopes, Bulgaria.* Microporous and Mesoporous Materials, 2010. 130(1): p. 32-37.
[19] Akhigbe, L., S. Ouki, and D. Saroj, *Disinfection and removal performance for Escherichia coli and heavy metals by silver-modified zeolite in a fixed bed column.* Chemical Engineering Journal, 2016. 295: p. 92-98.
[20] Bo, A., et al., *Removal of radioactive iodine from water using Ag2O grafted titanate nanolamina as efficient adsorbent.* Journal of Hazardous Materials, 2013. 246-247: p. 199-205.
[21] Mulfinger, L., et al., *Synthesis and Study of Silver Nanoparticles.* Journal of Chemical Education, 2007. 84(2): p. 322.
[22] Yoshida, S., Y. Muramatsu, and S. Uchida, *Studies on the sorption of I− (iodide) and IO3− (iodate) onto Andosols.* Water, Air, and Soil Pollution, 1992. 63(3): p. 321-329.

Acknowledgments

The financial assistance of the European Commission, Horizon 2020 MSCA-RISE-2016 project "NanoMed" towards this research is hereby acknowledged.

IOP Conf. Series: Earth and Environmental Science **182** (2018) 012015 doi:10.1088/1755-1315/182/1/012015

Study of Molybdenum Extraction from Alkali Roasted and Water Leaching of Ferro-Molybdenum Slag by Using TOA and TBP

Wei-Sheng Chen[1], Wen-Cheng Liang[1] and Cheng-Han Lee[1]

1 Department of Resources Engineering, National Cheng Kung University, Taiwan
E-mail: sky940412@gmail.com

Abstract. Solvent extraction of molybdenum (VI) from alkali roasted and water leaching of ferro-molybdenum slag by using TOA and TBP as extractants was investigated. In this study, ferro-molybdenum slags contain about 1 wt% Mo. Before Mo solvent extraction, Mo slags were roasted by sodium hydroxide under 6:1 NaOH–slag mass ratio and 600 ◦C for 2hr produced 99.5% molybdenum leaching efficiency by hot water. And leave impurity metal like Fe, Mg and Ca ions in leaching Raffinate. In solvent experiments, the organic phase is composed of TOA as extractant, TBP as modifier and kerosene as diluent. The effect of solvent extraction efficiency on different parameters such as pH and TOA/TBP concentrations. The result indicated the process was an effective method to separate molybdenum ions from high-impurity ferro-molybdenum slag.

1. Introduction

Molybdenum is an essential strategic metal for the infrastructure. As the development of modern society, the ceaseless exploitation of molybdenum will be exhausted. It's predicted that exhaustion of molybdenum within fifty to hundred years and there is little or no substitution potential for molybdenum in its major applications; however, the current rate of recycling molybdenum is only 20%. Therefore, it's important to increase the recycling rate of molybdenum from end-of-life products and develop effective methods to recycle low grade molybdenum and secondary molybdenum resources on hydrometallurgical processes [1].

The major application (more than 80%) of molybdenum is in steel and alloy usages. To produce high quality steels, ferro-molybdenum is a necessary additive. There are different molybdenum compounds to produce ferromolybdenum [2], such as calcium molybdate ($CaMoO_4$) and molybdenite using combustion synthesis method**,** a mixture of molybdenum sulfide, iron and carbon in sodium carbonate using the thermal dissociation method**,** taking molybdenite as raw material in lime as desulfurization reagent and molybdenite concentrate in the mixture of aluminum.

In the ferromolybdenum manufacturing process, MoO_3 is mixed with silicon, iron, aluminum and other metals and then reduced by an aluminothermic reaction. Therefore, a great number by-product of ferro-molybdenum slags were generated. The slags are composed of molybdenum, silicon, iron, aluminum, calcium, magnesium oxide and so on kinds of metal oxides. In this study, ferro-molybdenum slags include about 1 wt% molybdenum (The molybdenum content of viable ore bodies ranges between 0.01 and 0.25 wt%.). Research has therefore been initiated into treating ferro-molybdenum slags by direct metallurgical processes.

The purpose of the present paper is to develop a metallurgical process for treatment of the low-grade molybdenum from ferromolybdenum slags. To utilize molybdenum from ferro-molybdenum slag, alkaline roasting is an effective method to convert slags to sodium salts, such as sodium molybdate [3], sodium silicate and sodium aluminate [4], which can be almost decomposed by water leaching. In this study, water leaching is also good to get optimum molybdenum leaching efficiency for low-grade molybdenum in slag.

The polymeric state of molybdenum(VI) anions depends on the aqueous acidity and the species of mineral acid. The existence of molybdenum species in aqueous solution increases in the order $MoO4^{2-}$, $Mo_7O_{24}^{6-}$, $Mo_8O_{26}^{4-}$ and $MoO2^{2+}$ as the acidity of aqueous solution increases [5]. Different pH values form various types of ions or cations and influence extractants working.

Depending on different acid aqueous media affects the kind of extractant, including sulfuric acid by Alamine 304-I [6], PC-88A [7] and LIX®63[8], hydrochloric acid by TOPO and Alamine-308[9] and by TOA and TBP [10]. Although effective separation of molybdenum from acid solutions using the amine extractant of TOA has been reported [10], there is no consideration other impurity metal ions in leach solution. Therefore, the focus of the present paper is to investigate the optimum parameters through solvent extraction from high-impurity molybdenum water leach solution. This study investigated the direct extraction of molybdenum from water leach solutions containing molybdenum, silicon, aluminum and so on kinds of metal ions using an amine extractant of TOA and modified agent as TBP.

2. Experimental and Method

2.1. Materials, reagents and instruments

The samples of ferro-molybdenum slag were used for the experiments. The semi-quantitation analysis was conducted by X-ray fluorescence analyzer (XRF, Spectro XEPOS). The chemical composition (presented in Table 1.) was analyzed by inductively coupled plasma optical emission spectrometry (ICP-OES, Varian, Vista-MPX) with dissolving in lithium metaborate($BLiO_2$) under operational conditions $BLiO_2$/Slag ratio of 8/1, melting at 900°C for 2hr, and followed by hydrochloric leaching.

The aqueous solutions of molybdenum were prepared by leaching alkali roasted ferro-molybdenum slag using water. Mo leaching efficiency can get to over 99.5% under the optimum alkali roasting conditions that NaOH/slag ratio, roasting temperature, roasting time are 6g:1g, 600°C, 2hr, respectively and under the optimum water leaching conditions that leaching liquid-to-solid ratio, leaching temperature and leaching time are 150, 60°C and 60min, respectively. This water leaching procedure is effectively to leave Fe, Mg and Ca in the leaching raffinate, so the leaching solution contained molybdenum and other metal ions were obtained, the concentration of Mo, Si, Ca, Fe, Mg, Al, Na, K were presented in Table 2. by inductively coupled plasma optical emission spectrometry (ICP-OES, Varian, Vista-MPX). The water leaching solutions prepared as mentioned above were used as the feed solution and were adjusted pH before extraction using sulfuric acid and sodium hydroxide in the experiments. The leach solution had high concentrations of impurity metals (especially Si and Al) where molybdenum existed in the state of Mo(VI).

Analytical grade reagents, Sulfuric acid and sodium hydroxide, obtained from Sigma-Aldrich, were used for adjusting pH and roasting, representatively. Ammonium hydroxide, obtained from Honeywell, was used as a stripping reagent. Commercial kerosene, obtained from CPC Corporation, was used as a diluent to prepare the organic phase. Trioctylamine (TOA), an anionic extractant, obtained from Acros Organic, was used for the extraction of molybdenum. Tributyl phosphate (TBP), obtained from Alfa Aesar, was used as a modifier. All chemicals were used as-received without any further purification. All aqueous solutions were prepared using distilled water.

2.2. Experimental procedure

In solvent extraction experiments, 20 mL of leach solutions (aqueous phase) were mixed with different volumes of extractant (organic phase) of a desired concentration in kerosene diluent added to separating funnels at room temperature. To adjust the aqueous solutions pH were using 18 mol/L H_2SO_4 or 10 mol/L NaOH.

Table 1. Composition of ferro-molybdenum (wt %)

MoO_3	SiO_2	Al_2O_3	Fe_2O_3	CaO	MgO	Na_2O	K_2O	other
1.85%	48.80%	26.28%	14.03%	2.95%	3.30%	1.43%	0.73%	0.62%

Table 2. Composition of water leach solution (ppm)

Mo	Si	Al	Fe	Ca	Mg	Na	K	other
50	930	353	-	-	-	1830	36.7	N/A

Solvent extraction experiments were carried out thoroughly mixed at room temperature by mechanically shaking and after several minutes delaminated phases were separated. Then organic phases were collected (for stripping studies), and aqueous phases were analyzed, using ICP-OES, for molybdenum and other metal ion content. To find out the optimum condition of Mo extraction efficiency, adjusting different parameters such as leach solution pH, TBP/TOA ratio and concentrations, A/O ratio and contact time.

The distribution ratio D, according to eqs. (1), was calculated as concentration of metal present in the organic phase to that part in the aqueous phase at equilibrium. The extraction percentage (E%) was calculated as the mass of metal extracted into the organic phase to the initial mass of metal in the aqueous phase before extraction by eqs. (2), where V_{aq} and V_{org} are the volumes of aqueous and organic phases, respectively.

$$D = \frac{[M]_{org}}{[M]_{aq}} = \frac{C_i - C_f}{C_f} \times \frac{V_{aq}}{V_{org}} \quad (1)$$

$$E\% = \frac{D}{D + \frac{V_{(aq)}}{V_{(org)}}} \times 100 \quad (2)$$

Stripping experiments were conducted using the loaded organic solutions and strip solutions of optimum concentration of NH_4OH and O/A ratios thoroughly mixed for several times at room temperature by mechanically shaking.

The recommendation flow sheet of recovery molybdenum from ferromolybdenum slag as **Fig.** 1.

2.3. *Analytical methods*

The metal contents of sample and leaching residues were precisely determined by inductively coupled plasma optical emission spectrometry (ICP-OES, Varian, Vista-MPX).

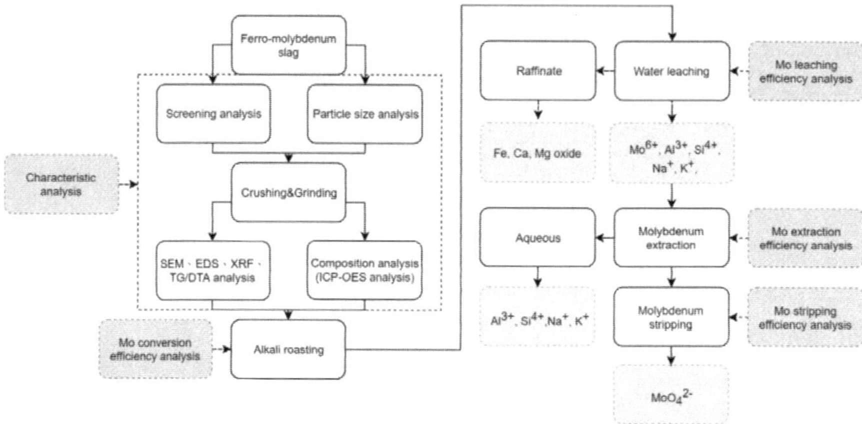

Figure 1. The flow sheet of recovery Mo from molybdenum slag

3. Results and Discussion

3.1. Extraction efficiency of molybdenum

The extraction reaction of molybdenum in sulfuric acid media with TOA steps as follow: (i) To form the hydroacid of ammine by acid transfer, (ii) ion exchange of anionic molybdenum complexes, according to Eqs. (3) and (4). And the overall reaction can be described by Eq. (5).

$$\overline{R_3N_{(org)}} + H_2SO_{4(aq)} = \overline{R_3NH \cdot HSO_{4(org)}} \tag{3}$$

$$\overline{2R_3NH^+ \cdot HSO^-_{4(org)}} + MoO^{2-}_{4(aq)} = \overline{(R_3NH)_2 \cdot MoO_{4(org)}} + 2HSO^-_{4(aq)} \tag{4}$$

$$\overline{2R_3N_{(org)}} + 2H^+_{(aq)} + MoO^{2-}_{4(aq)} = \overline{(R_3NH)_2 \cdot MoO_{4(org)}} \tag{5}$$

3.1.1. Effect of pH on molybdenum extraction efficiency

The effect of pH on molybdenum extraction efficiency was investigated. The water leach solution was adjusted pH from 1 to 7. In this experiment, using TOA=0.05M, TBP/TOA=3(v/v), A/O=5:1, and contact time=10min. The result was shown in **Fig. 2**. As we knew the eq. (3), it was necessary to have enough H^+ to form the hydroacid of ammine, so there was no reaction when the pH was about 5-7. As the acidy increased and pH decreased, Mo extraction efficiency increased rapid. When the pH was 2 to 3, there was a good extraction efficiency. Furthermore, when pH was equal to 2.6, it was the optimal parameter to have the high molybdenum extraction efficiency according the experiment. And there was few amount of co-extraction of Si and Al along with molybdenum in the above pH range.

3.1.2. Effect of TOA and TBP concentrations on molybdenum extraction efficiency

The effect of TOA concentrations varied from 0.01M to 0.1M on molybdenum extraction efficiency was shown in **Fig. 3**. TOA concentration from 0.01M to 0.05M with extraction efficiency of molybdenum slowly increased from 97.8% to 99.5%. The high extraction efficiency of Mo was under 0.05M. However, it was observed that a slight emulsion was shown during the extraction. Therefore, TBP is as a modifier agent to improve the organic solution [10]. The effect of TBP/TOA ratio influenced the observation of phase separation. As the Table 3. When TBP/TOA = 1or2, third phase was formatted. As TBP/TOA was over 3, phase separation was shown up. The ratio of TBP/TOA was kept constant at 3 to inhibit formation of emulsion or third phase [11]. As the result observed above, to maintain high molybdenum extraction efficiency and improve phase separation, the optimum parameter of TOA and TBP concentrations are 0.05M and TBP/TOA=3(v/v).

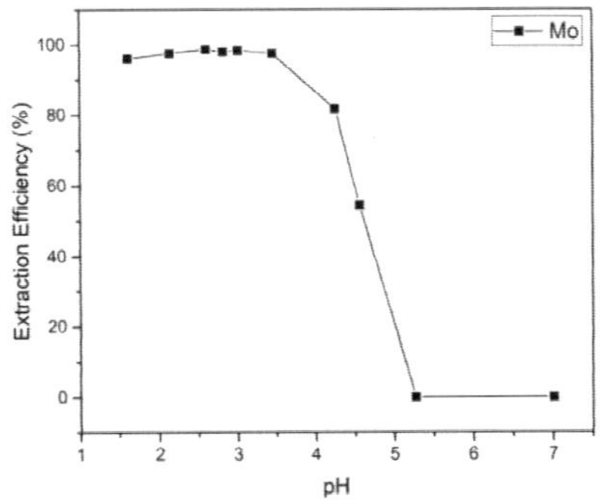

Figure 2. The effect of pH

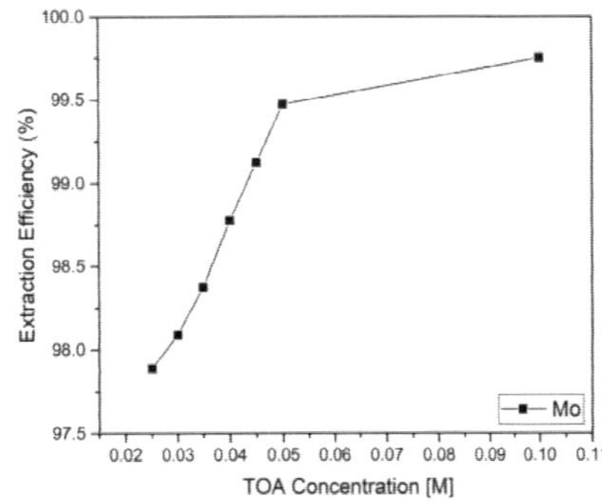

Figure 3. The effect of TOA concentration on extraction

Table 3. Effect of TBP/TOA on phase separation

SL. No	TBP/TOA	Observations
1	5%/5%=1	Third phase formation
2	10%/5%=2	Third phase formation
3	15%/5%=3	Two phase separation
4	20%/5%=4	Two phase separation
5	25%/5%=5	Two phase separation

3.1.3. Effect of A/O ratio on molybdenum extraction efficiency

The effect of A/O ratio varied from 0 to 20 on molybdenum extraction efficiency was shown in **Fig. 4**. When the A/O was from 1 to 5, it was getting high Mo extraction efficiency up to 99.5% under the condition of TOA of 0.05M, TBP/TOA of 3 and time of 10mins. Although the A/O ratio increase makes the efficiency decrease slowly, A/O ratio can concentrate the concentration of Mo. Therefore, choosing the optimal A/O=5 was the optimal parameter.

3.1.4. Effect of extraction time on molybdenum extraction efficiency

Extraction time also influenced the extraction efficiency. As the **Fig. 5**, discussing the effect of extraction time on Mo extraction efficiency under the condition of TOA of 0.05M, TBP/TOA of 3 and A/O of 5 . When the extraction time is 1 min, the efficiency is about 96%. As the time got to 3 mins, it was close to achieve 99% efficiency. Furthermore, 5 mins had over 99.5% efficiency. To ensure the efficiency of Mo extraction, selecting 10 mins was as the optimal time parameter.

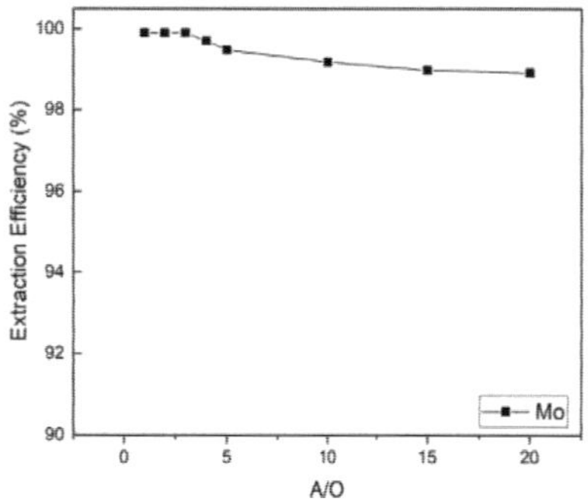

Figure 4. The effect of A/O ratio on extraction

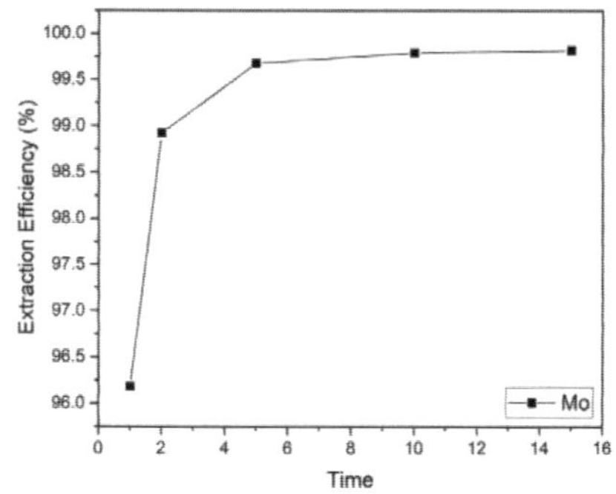

Figure 5. The effect of time on extraction

3.2. Stripping efficiency of molybdenum

In this study, the loaded organic contained about 50ppm Mo. To recovery Mo from loaded organic, it was necessary to determine the optimal parameter of stripping experiment, such as the concentration of stripping reagent, stripping O/A and stripping time. The stripping reagent was NH_4OH which stripped Mo from the organic phase using this research. The loaded organic contained Mo could be stripped from ammonia solution because of the dissociation of the extracted complex at alkaline conditions. The stripping reaction is presented as below eqs. (6).

$$\overline{(R_3NH)_2 \cdot MoO_{4(org)}} + 2NH_4OH_{(aq)} = \overline{2R_3N_{(org)}} + (NH_4)_2MoO_{4(aq)} + 2H_2O \quad (6)$$

3.2.1 Effect of ammonium hydroxide concentration

According to the amount Mo contained in organic phase, to select the appropriate amount of ammonium hydroxide. The NH_4OH concentration from 0.005 to 0.030 M at an O/A ratio of 2 and time of 10mins. The result was presented as **Fig. 6**. When the concentration of NH_4OH was 0.030M, the stripping efficiency of Mo was getting to 99.9%.

3.2.2 Effect of O/A ratio

Under the stripping condition of NH4OH was 0.030M and the time was 10mins. To discuss the influence of the O/A ratio from 1 to 3. The Mo extraction efficiency achieved 99.9% at the O/A of 1 and kept over 99.5% at the O/A of 2. Within the O/A of 3, Mo extraction dropped to 93%. Therefore, the optimal A/O ratio kept at 2. As the result presented on **Fig. 7**.

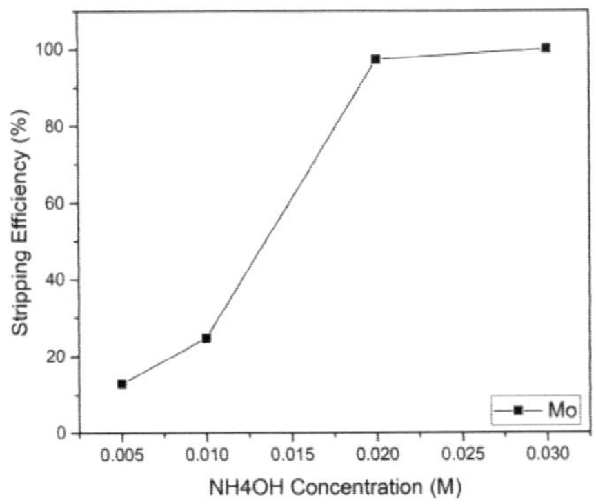

Figure 6. The effect of NH4OH concentration

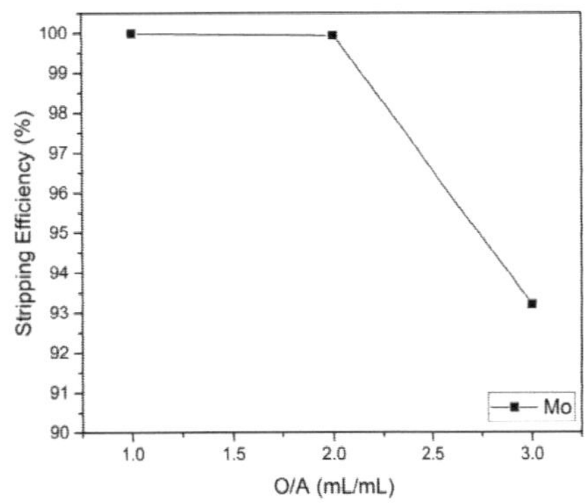

Figure 7. The effect of O/A ratio

4. Conclusions

The leach solution was produced by alkali roasting and then water leaching from ferro-molybdenum slag, which can make Fe, Ca and Mg impurity metal leave in the leach raffinate.

Under the optimum extraction condition of pH=2.6, A:O=5:1, TOA of 0.05M, TBP/TOA=3(v/v) and time=10mins, to get a high extraction efficiency of molybdenum. And there is no third phase formation and emulsification under optimum parameter. The optimal stripping parameter of NH4OH and O/A ratio was 0.03M and 2, respectively.

5.References

[1] M.L.C.M. Henckens, P.P.J. Driessen, E. Worrell, "Molybdenum resources: Their depletion and safeguarding for future generations," *Resources, Conservation and Recycling*, vol. 134, pp. 61-69, 2018.

[2] M.H. Golmakani, J. Vahdati khaki, A. Babakhani, "A novel method for direct fabrication of ferromolybdenum using molybdenite via self-propagation high temperature synthesis," *Materials Chemistry and Physics*, vol. 194, pp. 9-16, 2017.

[3] Lihua Shi, Xue-Wen Wang, Ming-Yu Wang, Jun Peng, Caixia Xiao, "Extraction of molybdenum from high-impurity ferromolybdenum by roasting with Na_2CO_3 and CaO and leaching with water," *Hydrometallurgy*, vol. 108, pp. 214-219, 2011.

[4] Desheng Chen, Longsheng Zhao, Yahui Liu, Tao Qi, Jianchong Wang, Lina Wang, "A novel process for recovery of iron, titanium, and vanadium from titanomagnetite concentrates: NaOH molten salt roasting and water leaching processes," *Journal of Hazardous Materials*, vol.244-245, pp. 588-595, 2013.

[5] Zhongwei Zhao, Liang Yang, Guangsheng Huo, Xingyu Chen, Haijun Huang," Solvent extraction of molybdenum blue from alkaline leaching solution of the Ni-Mo ore," *International Journal of Refractory Metals and Hard Materials*, vol. 29(2), pp.232-236, 2011

[6] P.K. Parhi, Kyung-Ho Park, Hong-In Kim, Jin-Tae Park, "Recovery of molybdenum from the sea nodule leach liquor by solvent extraction using Alamine 304-I," *Hydrometallurgy*, vol. 105, pp. 195-200, 2011

[7] Xia Yun, Xiao Liansheng, Xiao Chao, Zeng Li, "Direct solvent extraction of molybdenum(VI) from sulfuric acid leach solutions using PC-88A," *Hydrometallurgy*, vol. 158, pp. 114-118, 2015

[8] Zeng Li, Cheng Chu Yong, "Recovery of molybdenum and vanadium from synthetic sulphuric acid leach solutions of spent hydrodesulphurisation catalysts using solvent extraction," *Hydrometallurgy*, vol. 101, pp.141-147, 2010

[9] Banda Raju, Sohn Seong Ho, Lee Man Seung, "Process development for the separation and recovery of Mo and Co from chloride leach liquors of petroleum refining catalyst by solvent extraction," *Journal of Hazardous Materials*, vol. 213-214, pp.1-6, 2012

[10] Ghadiri Mehdi, Ashrafizadeh, Seyed Nezameddin, Taghizadeh Mohammad, "Study of molybdenum extraction by trioctylamine and tributylphosphate and stripping by ammonium solutions," *Hydrometallurgy*, vol.144-145, pp. 151-155, 2014

[11] Zhan-fang Cao, Hong Zhong, Zhao-hui Qiu, "Solvent extraction of rhenium from molybdenum in alkaline solution," Hydrometallurgy, vol.97, pp. 153–157, 2009

IOP Conf. Series: Earth and Environmental Science **182** (2018) 012016 doi:10.1088/1755-1315/182/1/012016

Preliminary Study on Thermodynamic Urban Design Based on Prototype Research of Tropical Rainforest

FAN Yating[1]

1 D.Arch. University of Hawaii at Manoa
E-mail: yatingf@hawaii.edu

Abstract. Under the background of contemporary global design and all-sided sustainable development, traditional architectural design methods and the autonomy of architecture are being challenged. Energy and thermodynamics now provides a more complete perspective on the future based on archeology and science. Before attempting to find a city paradigm for Singapore, a tropical island nation facing an energy crisis, this article, from the perspective of natural learning, examines the ecosystem of tropical rainforests, conducting the prototype analysis with four aspects, climate and microclimate, vertical structure, energy flow and ecological community. Correspondingly, four important strategies that run throughout the thermodynamic urban paradigm design are summarized: climate adaptation, self-organization, community system and regeneration succession. Afterwards to plus the further design from four different levels: region, city, architecture and experience, this article is to explore new thermodynamic urban paradigm.

1. Introduction

The research on "energy" has been the clue throughout natural science and social history since the recent half century. In the energy context, architecture can be considered as a kind of "open and unbalanced system". According to Prigogine's point view, it also can be deemed as the "dissipative structure" of thermodynamics. Therefore based on this perspective, architecture, a dissipative structure, is featured with energy exchange maximization and entropy maintenance, which would raise the repeated consideration on architecture based on the consideration on energy within the overall thermodynamic system. The process, extensive and sustainable development of globalization, is another background of current issue about the energy and thermodynamics, which puts forward new requirements for the autonomy and revaluation of architecture from the perspective of energy.[1-7]

2. Prototype research on tropical rainforest

In order to seek for a thermodynamic vertical city design strategy especially in tropic, this paper firstly turns to nature learns from nature to explore the organization form in the natural ecosystem which is different from the modernism Le Corbusier's Delirious New York. The tropical rainforest, as the ecosystem in the tropical zone, has survived and grown the natural selection for twenty million years, ancient and mysterious, efficient and friendly. Its open system and organization form definitely become the outstanding objects for prototype research by taking energy as the perspective.

2.1. Climate adaptation feature

2.1.1. Tropical climate response. Nearly all the features of the tropical rainforest ecosystem are formed due to the response to the tropical climate, in other words, the tropical rainforest is a climate-orientated ecosystem, and completely complies with the natural environment locality (figure 1). The climate features of tropical rainforest different from other ecosystems mainly include three following aspects. 1. The tropical rainforests are generally located between the 23.5 parallels enjoys perennial direct sunlight, which provides the possibility of severe photosynthesis for rainforest and provides prerequisite for its biomass up to 2.9kg/m2*y. 2. The rainfall is extremely abundant, basically higher than 1800mm and even reaching to 3500mm, which no doubt intensifies the respiratory and transpiration of plants, heavily promoting the metabolism; 3. Keeping at high temperature level without cold winter and obvious diurnal temperature variation, therefore, makes tropical rainforest always in the energy circulation of producing, consuming and reproducing.[7,8]

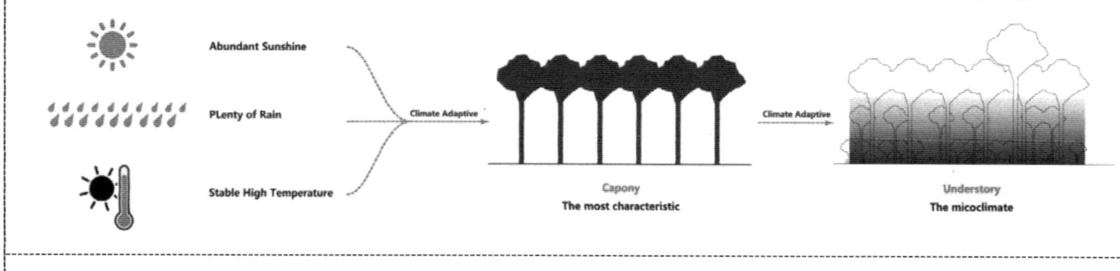

Figure 1. Climate orientation [a]

[a] drawn by the author

2.1.2. Microenvironment and microclimate establishment. Under the environment of sufficient direct sunlight, abundant rainfall and constant high temperature, the crowns are all striving for growth to enjoy more sunlight and rainfall. So that the canopy forms a continuous expansion surface to blot out the sky like a cove. At that time, the dwarf tree, other dwarf tree species, shrubs and mosses as well as almost all animals completely live in an environment different from the canopy expansion surface. In other words, the canopy establishes a microenvironment and microclimate for the understory part (figure 2). The characteristics are as follows. 1. Breezy environment. 2. Weak light environment. The canopy blocks almost all sunlight. 3. Light quality. The light spectrum penetrating into the bottom environment cannot promote blossom and fruit yielding. 4. Sharp increase in humidity. 5. Interception of Rainfall. 6. Weak temperature variation. [7,8]

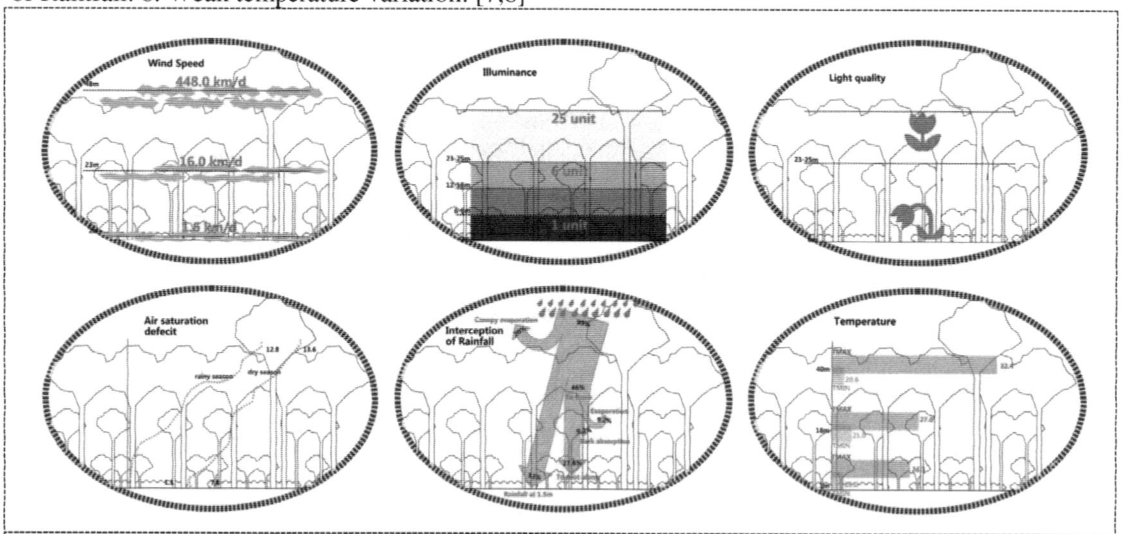

Figure 2. Undergrowth microenvironment and microclimate [a]

[a] drawn by the author

2.2. Energy flow feature

Among the terrestrial ecosystems, the tropical rainforest is no doubt the most productive ecosystem with the net primary productive force of 2.9kg/m^2*y, compared with 1.5 kg/m^2*y for the agriculture in temperate regions, 1 kg/m^2*y for the tropical grassland, even negative for the ocean and land ecosystems. The primary production of tropical rainforest accounts for 13.76% in the earth, while it reaches to 40.4% in the total biomass. [7,8]

2.2.1. Energy capturing and guidance of a big tree. One adult tree in the tropical rainforest can produce pure sugar of 1.5kg within one day, where the direct breathing of plants and life consumption account for 60%, while the left 40% are used for chemical matrix synthesis, such as the vegetable protein and fiber for wood, and only 10% are converted to secondary productivity, namely, the protein eaten by animals. Complex carbon molecules, namely, the organic matters retained in the animal waste and corpses of animals and plants, fall onto the ground to provide food for fungus and germ which decompose and restore them (figure 3). There is a mystical phenomenon in the tropical rainforest, which is the soil is relatively infertile. It is the fast energy circulation that causes the thin humus layer of soil. Therefor the tree roots rely on fungus to absorb and circulate energy, that is to say, a kind of symbiotic relationship, mycorrhiza. Hyphae makes the dead organic substances to be absorbed by tree roots, while only small part of organic substances are decomposed, restored to be soluble nutrient substance by fungus to enter into soil and absorbed by plants. [7,8]

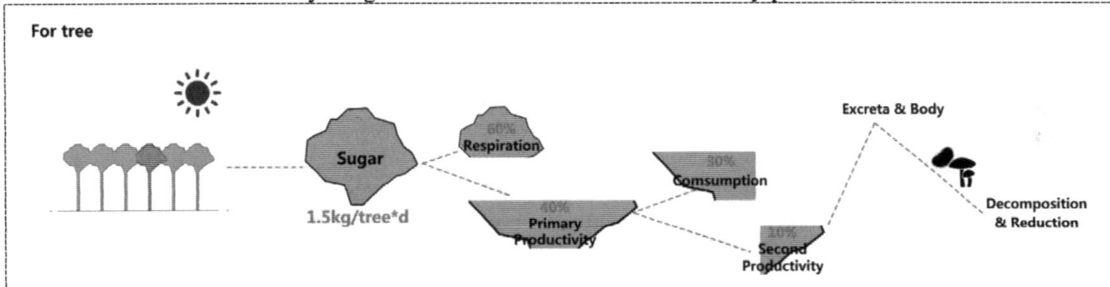

Figure 3. Energy capturing and guidance of a big tree [a]

[a] drawn by the author

2.2.2. Energy flow of the rainforest. The 4 000 000 of the 5 000 000 units solar energy are absorbed and reflected by clouds with only 1 000 000 shine left at the canopy layer. Wherein, half of the energy are absorbed and converted by the canopy and even the leaf with only 2000 units energy are absorbed by phytophagous animals. Through the stage shedding of biological food chain, the final 40 units energy can be fallen to the ground, converted or absorbed by fungus. The energy is being constantly dissipated in the energy flow process (figure 4). [7,8]

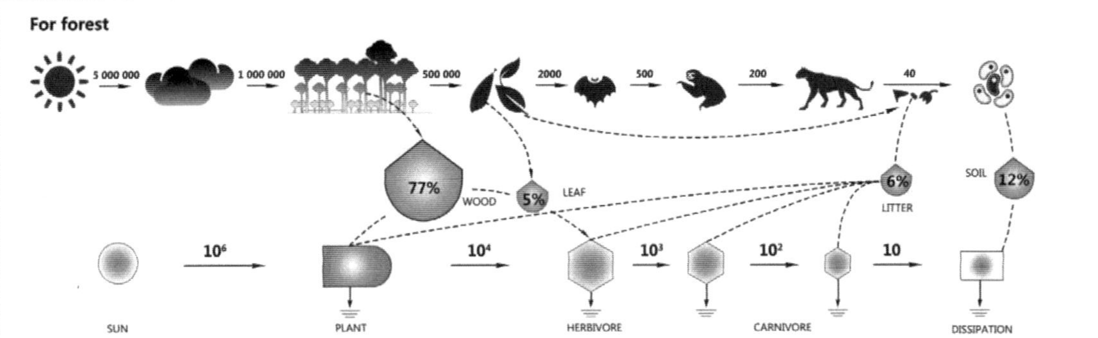

Figure 4. Energy flow of the rainforest [a]

[a] drawn by the author

2.3. Vertical Spatial Heterogeneity

IOP Conf. Series: Earth and Environmental Science **182** (2018) 012016 doi:10.1088/1755-1315/182/1/012016

Due to climatic reasons, tropical rainforest forms a thick layer of continuous canopy, which then creates a sub-forest micro-environment with gradient differences. Furthermore, this micro-climate, in turn, makes the under-forest environment vertically layered, forming a unique spatial heterogeneity in the tropical rainforest (figure 5). [7,8]

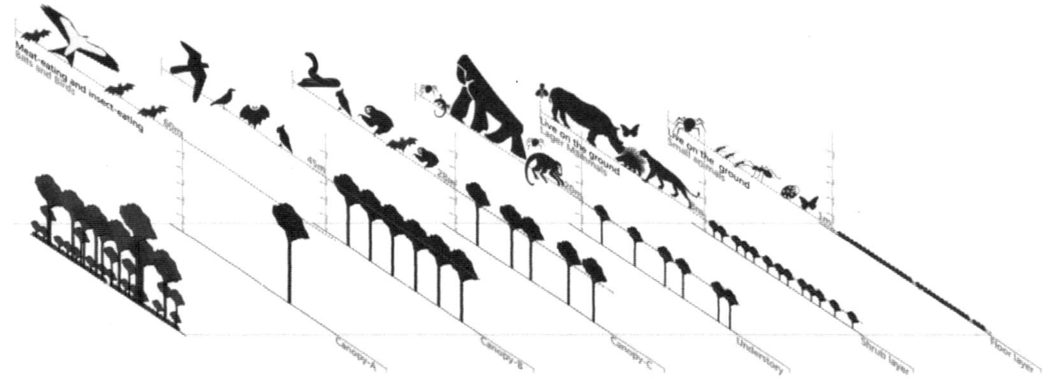

Figure 5. Vertical spatial heterogeneity of flora and fauna [a]

[a] drawn by the author

2.3.1. Overview of flora vertical layering. There are vertical stratifications in the tropical rainforest, including the canopy-A layer, the canopy-B layer, the canopy-C layer, the understory leaf layer, the shrub layer and the floor layer. Among them, the canopy-A layer refers to the part outside the edge of canopy, which is the highest canopy, even reaching to 84 meters in South America. The canopy-B and the canopy-C layers are closely spaced trees and their ripe dense top crowns. Understory leaf layer refers to the more widely separated, smaller tree species and young plants that form the fracture layer under the canopy. The shallow shrub layer consists only of shrubs and young trees growing less than 10 meters above the ground. The floor layer includes tree seedlings, fungi, and low-growth plants. The vertical structure of flora divided into 6 layers, nonetheless, is not always obvious for the vertical structure will change due to differences in climate, soil and other factors. [7,8]

2.3.2. Overview of fauna vertical layering. Corresponding to the phenomenon of six levels of flora stratification, Harrison (1962) believed that in the rainforest, the fauna vertical stratification seems to be more apparent. He also summarized and classified the animal's vertical stratification into six levels. 1. The space above the canopy, that is, the canopy at low altitude, mainly including insectivores and carnivorous animals, dominated by bats and large birds. 2. Both in the canopy-A and B layers, there are herbivorous mammals, as well as insectivorous animals and omnivores, such as various types of birds and fruit-eating bats. 3. Under the canopy, there are mainly flying animals and insectivorous bats in the middle area between the trunks. 4. The space full of rattan vines on the trunks, there are climbing carnivorous animals and insects that feed on the epiphytes on the trunks, such as the big spider. 5. On the ground, there are large mammals who can run. 6. And small ground animals. Overall, more than half of the mammals inhabit the canopy, and 70%- 80% of these lives live on trees. [7,8]

2.3.3. Causes and significance of spatial heterogeneity. In an ecosystem, the position occupied by time and space and its functional relationship with related populations are the niche. When there are spatial differences in resources, and moreover each species settles in a position where it has a competitive advantage, there is a possibility that the species will coexist. In tropical rainforest, the canopy shields the undergrowth environment like a shell, which forms a micro-environment with a gradient that allows the plants to be vertically stratified. As a result, the animals that depend on it also vertically occupy different niches. In this way, plants and animals in the ecological niche with advantageous differences in time and space are continuously producing new species. Therefore, a huge variety of species lives together in the rainforests. In other word, the spatial heterogeneity has successfully allowed these diverse species to co-exist. [7,8]

2.4. Ecological communities symbiotic

2.4.1. Overview of community diversity. The biological species of tropical rainforests account for 70% of the earth. The study found that such a large number of organisms survived in the tropical rainforest with extremely rich diversity. The tropical rainforest that developed from the age of dinosaurs 20 million years ago has 30,000 advanced plant species. However, the competition among rich species is relatively weak, and thus it is possible to coexist with each other. [7,8]

2.4.2. Four hypotheses of causes for the coexistence of various species. How do such rich and diverse species form and why do they coexist? There are four hypotheses (Figure 6). 1. Niche Differentiation hypothesis is that when there are spatial differences in resources, and each species settles in the position with its competitive advantage, a large number of species can coexist. As the rainfall in the tropical rainforest increases, the plant species density increases, soil fertility decreases, spatial heterogeneity increases, and diversity increases. 2. Crazing or Pest Pressure hypothesis means that if there is a high probability of being fed and destroyed for species rich in diversity or competitiveness, these harmful feedings will promote the coexistence of multiple species. Predators control the structure of the community through damaging to the prey. 3. Life History Trade Off hypothesis means that negative correlations between abilities (communicative ability, competitive ability, and growth ability after settlement) may promote the coexistence of multiple species. If this assumption is strictly followed, then infinite species can coexist. 4. Lottery Competition hypothesis refers that on the one hand, most plants in tropical rainforests occupy the same ecological niche and have the same ecological needs as the ecological equivalent species. On the other hand, even if the differences in the competitive ability between species are not obvious, the chance of competition is small and there is no competition exclusion, then species can coexist. [7,8]

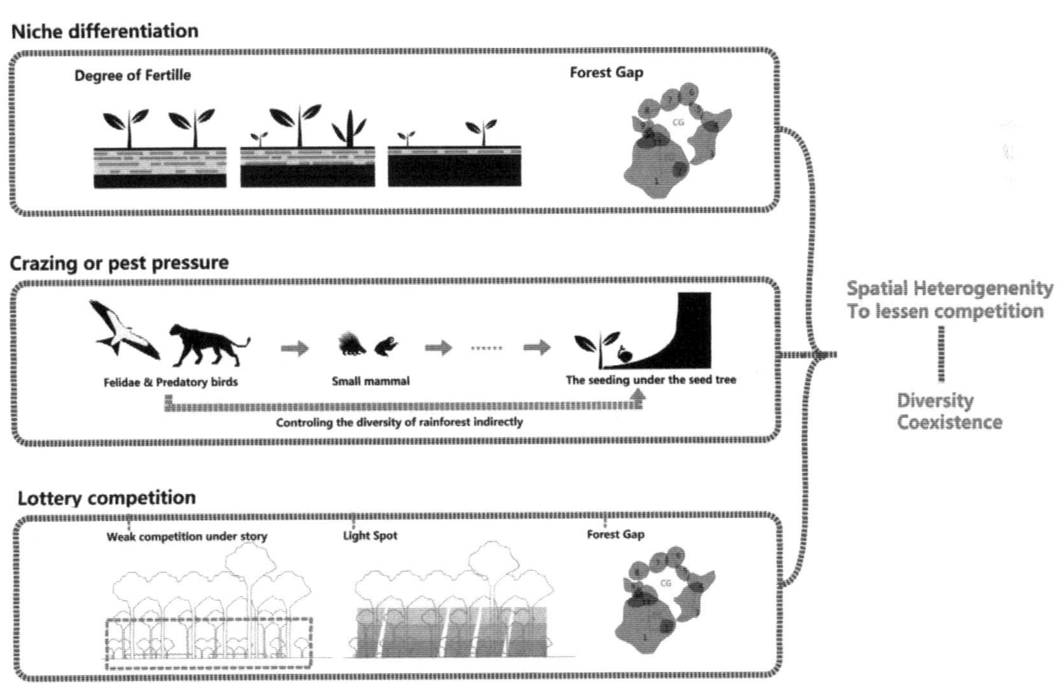

Figure 6. Hypothesis of Diversity Cause [a]

[a] drawn by the author

IOP Conf. Series: Earth and Environmental Science **182** (2018) 012016 doi:10.1088/1755-1315/182/1/012016

2.4.3. Growth and metabolism. The Lottery Competition hypothesis points out an interesting phenomenon - Forest Gap. Because of the existence of Forest Gap, it is randomness rather than competitiveness that determines which species can be successfully established under the forest. The tropical rainforest experienced the periods of pioneering succession, early secondary succession, late secondary succession and current climax succession. Lottery competition caused by forest gaps, etc. not only promotes the establishment of biodiversity, but also brings the tropical rainforest a stable and benign metabolism during its growth. [7,8]

3. Thermodynamic urban design strategy based on Singapore

Singapore is a city-state walking in the forefront of the world. In the face of various crises in the development of globalization today, although Singapore is an island nation, it has contributed a lot to the strategy of solving various crises. In the current global context of sustainable development, we should pay more attention to the flow of energy under the order of things. As the biologist, Odom said, energy flow can be extended to the environment, energy in society can be used to shape the landscape of culture and thought. In this context, we propose a prototype of a rainforest thermodynamic city to create an urban paradigm that not only contributes to Singapore but also as a solution facing the current global energy and environment crisis especially for cities in Asia that are rapidly developing. Therefore, based on the four important characteristics of the research on tropical rainforest prototype, we accordingly propose four important strategies for the tropical rainforest thermodynamics urban prototype.[1,7]

3.1. Climate response

The energy is the deep level order for the organization of things and the energy is closely related to the climate. The growth and four characteristics of tropical rainforests are all due to the climate oriented. The biosphere uses the most basic principle - climate adaptation, to shape the ecological system that with regional differences, and shaped the mysterious and ancient ecosystem like the tropical rainforest. For tropical cities, like Singapore, the climate response from the urban perspective, on the one hand, is reflected in the utilization and enrichment of the forest to form a green lung system for the city, making the overall effect is greater than the local sum. On the other hand, the design of the road system conforms to the main wind direction to form wind tunnels for the city, so as to capture the energy of the forest and water resources with the maximum efficiency. From the point of view of architecture, the design idea that changed from the design of space to the "guidance" of the energy in the sunshine, rain and wind, moreover to some extent the air has become the leading role in the design (figure 7 and 8).[4-6]

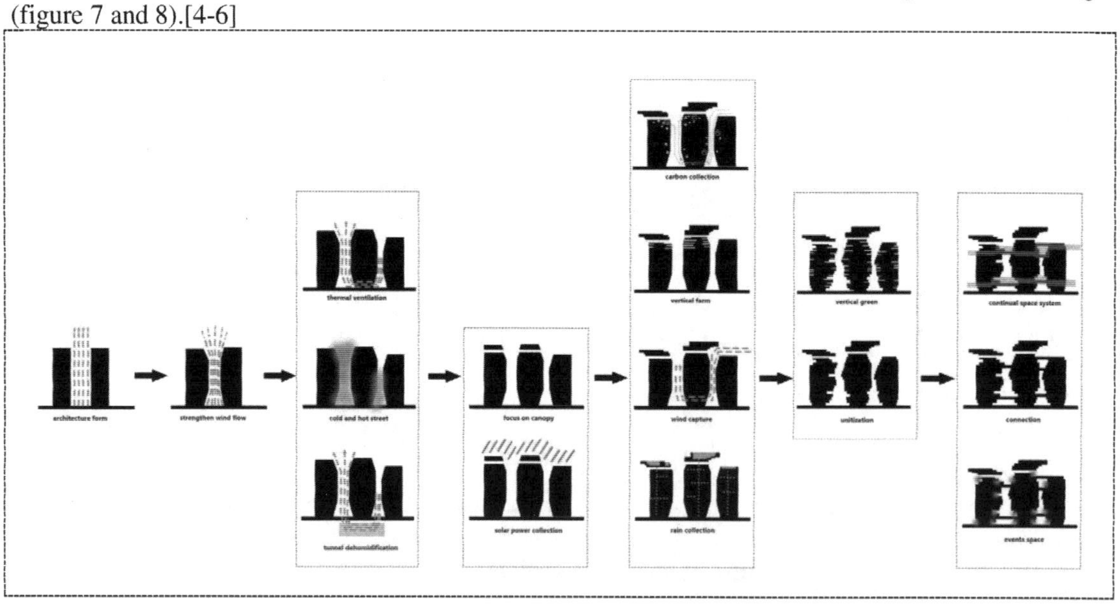

Figure 7. Formal following energy on the architectural scale [a]

[a] drawn by the author

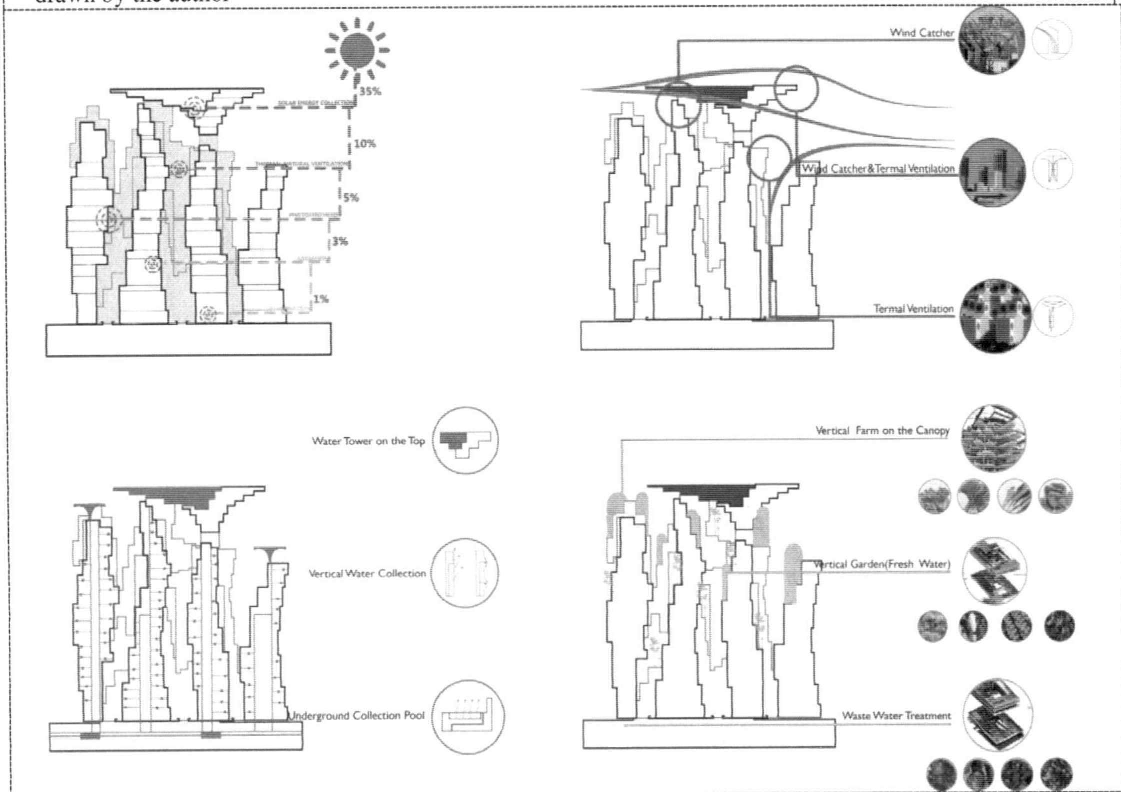

Figure 8. Energy capture on the architectural scale [a]

[a] drawn by the author

3.2. Self-organization

In its own evolution process, there is no need for the specific interference of the outside world. An open and unbalanced system flows and transforms the energy at the fastest speed, with the strongest feedback and the maximum amount. It can achieve a certain goal by the mutual coordination of its own elements, that is, that is to say, forming the Self-organization of the system. Self-organization is a way to understand and interpret the world in the context of complex science, also considered as an ideal mathematical and physical model. The energy flow of the tropical rainforest already approaches the self-organization to a large extent. How to let the operational pattern of the thermodynamic city model to fit the law of energy flow, the "hexagonal" structure that can be seen everywhere in the tropical rainforest even the whole nature has inspired us, together with the research of the architect Otto also concluded, that the hexagon is the most efficient and intensive structure for the boundary of a system and the core service range. Using this kind of organization mode in the thermodynamic urban design, on the one hand, can arrange and organize the social activities such as the road network and industries of the city, which makes energy circulate and transforming more sufficiently and conveniently between each system and reduces the energy loss caused by traffic and waiting time. On the other hand, it can pave the infrastructure deep into the community and even family households, such as the distributed smart power grid, distributed sewage treatment and biogas power generation, distributed rainwater collection based on new energy, to utilize the wind, rainwater, solar energy and even garbage to the maximum efficiency, thereby to complete the "self-sufficient" energy flow (figure 9 and 10). [9]

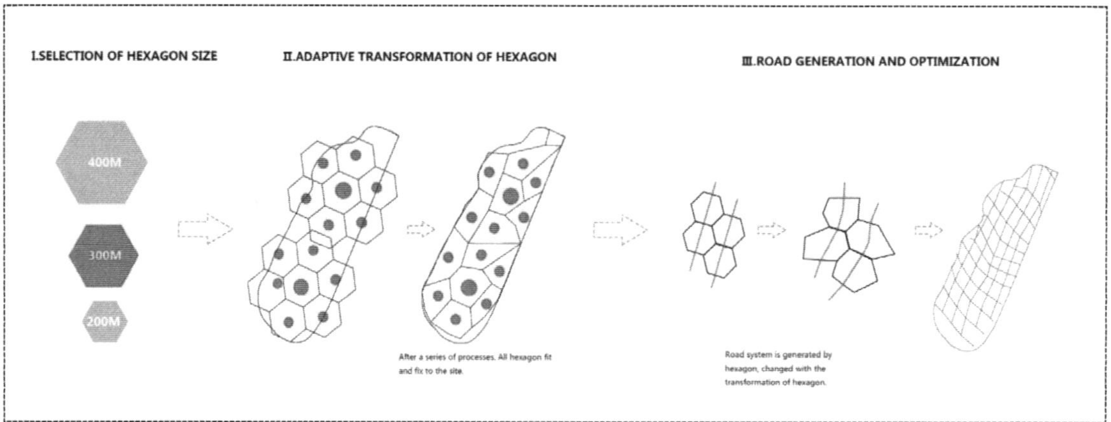

Figure 9. The "hexagon" structure acts on the urban form [a]

[a] drawn by the author

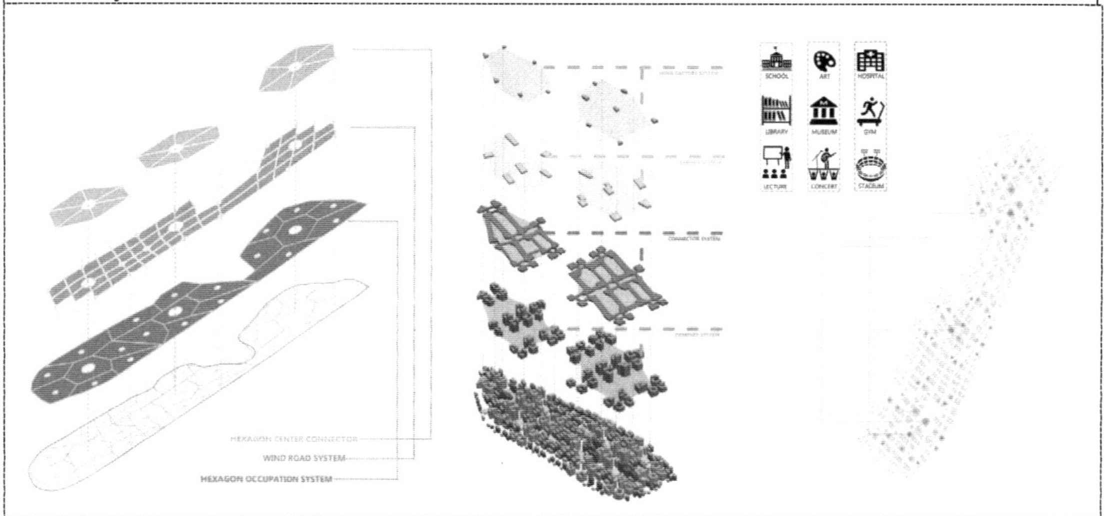

Figure 10. Distributed infrastructure system [a]

[a] drawn by the author

3.3. Community System

In the rainforest system, it is possible to say that the spatial heterogeneity of vertical stratification directly leads to the extremely abundant biodiversity, while the thermodynamic urban model should also encourage the sufficient community diversity as well. It is neither enough nor realistic to avoid competing for the society and state only directly by letting different communities occupy different time and space resources, and even by the negative correlation phenomena of life history. However when the different communities of the tropical rainforest occupy the common resources, different rules of symbiosis can be cultivated, like epiphytism, saprophytism, mutualism and killing, among which there are diverse forms of mutual benefit and coexistence models. For example in the relative barren soil humus of the tropical rainforest, mycorrhiza makes the root cells to absorb organic matter directly, thus the fungi gain the nutrition in this process as well. As an island state, Singapore possesses limited resources, and different ethnic groups and communities should find the living way of mutualism, which is the maximum utilization of resource energy efficiency. In the thermodynamic city paradigm, the canopy of high density vertical cities is responsible for capturing the energy and resource such as the sunlight, water and wind. Just like the dwarf tree layer and the tree trunk layer, the streets and the vertical public space of the city are the main carrier of urban life. Overall the different communities

IOP Conf. Series: Earth and Environmental Science **182** (2018) 012016 doi:10.1088/1755-1315/182/1/012016

occupy the suitable resources to live and work in the city, meanwhile to interact and communicate sufficiently in the common complex system (figure 11).[10,11]

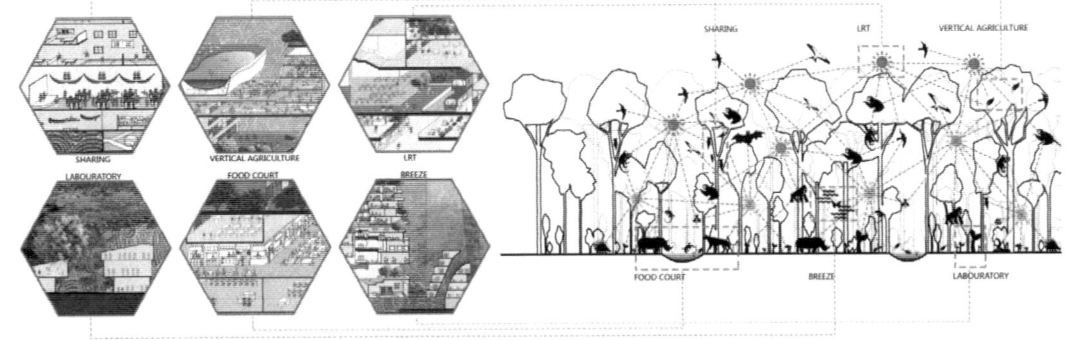

Figure 11. The urban community in the tropical rainforest [a]

[a] drawn by the author

3.4. Regeneration and Succession

The same as the rainforest system, the city is not a certain state of time, but similar to an organism that undergoing metabolic process continuously. Urban morphology and social activity will be in different stage of regeneration. In the thermodynamic urban design, on the one hand, the city keeps growing, the stage planning and construction of the city, and the gradual and sustainable growth and succession of the industry and the living population. On the other hand, the urban canopy is abundant but will not shelter all the sunlight and sky but with the "Forest Gap" formed randomly, to provide possibility for the flexible activities on the ground and the vertical city.[8,10]

4. Conclusion

Indeed, the so called "city paradigm" is an ambitious conception, and the current four design strategies of the "thermodynamic city paradigm in tropical rainforest" cannot completely solve all the social problems of the city in the future. However, still, from the perspective of energy and learning from the nature and form the thriving and ancient ecological system of the tropical forest, thereby to put forward the design strategy of thermodynamic city will undoubtedly provide a new perspective to solve the current urban problems such as the urgent resource shortage and energy crisis.

5. References

[1] Latour, Bruno. *We have never been modern*. Harvard university press, 2012.

[2] LI Linxue. Knowledge, discourse and paradigm historical scenario and contemporary frontier of energy and thermodynamic architecture. *Time Architecture* 2015-2:10-16

[3] William W.BRAHAM, ZHANG Boyuan. Thermodynamic narratives. *Time Architecture* 2015-2:26-31

[4] Kleidon A, Lorenz R D. *Non-equilibrium Thermodynamics and the Production of Entropy*. Springer Berlin Heidelberg, 2005.

[5] Kiel MOE, CHEN Hao. The nonmodern struggle for maximum entropy. *Time Architecture* 2015-2:22-25

[6] Inaki ABALOS, ZHOU Jianjia. Interior sources and sink. *Time Architecture* 2015-2:17-21

[7] Odum, Eugene Pleasants, Howard T. Odum, and Joan Andrews. *Fundamentals of ecology*. Vol. 3. Philadelphia: Saunders, 1971.

[8] Richards, Paul Westmacott. *The tropical rainforest; an ecological study*. At The University Press; Cambridge, 1952.

[9] Portugali J. *Self-Organization and the City*. Berlin: Springer-Verlag , 2000.

[10] Population White Paper. A sustainable population for a dynamic Singapore.2013-1

[11] Netina Tan. Multiracialism and Politics of Regulating Ethnic Relations in Singapore. "Politics of Identity and Nationalism" Panel, 2013-6

A Study on Daylighting Design of Urban Mid-Rise Housing from the Perspective of Carbon Emission Reduction Effect: Shanghai, China

Y Huang[1], L Li[2] and C E Llewellyn[3]

1 Ph. D/Professor, College of Architecture and Urban Planning
Director of Academic Affairs, Tonji University, Shanghai, China
2 M.Arch Student, Tongji University, Shanghai, China
D.Arch Student, University of Hawaii at Manoa, Honolulu, USA
E-mail: liling8@hawaii.edu
3 Ph. D/Professor, University of Hawaii at Manoa, Honolulu, USA

Abstract. Under the circumstance of rapid development, the contradiction and balance between energy consumption, carbon emission and urban living environment are increasingly become one of the problems to be solved in contemporary China. Housing has demonstrated tremendous potential to play a major role in the reduction of carbon emission, to gain a balance between reducing carbon emission and meeting increasing demand. Good daylighting is irreplaceable in improving the quality of housing and meeting the daily physiological and psychological needs of the residents. Thus, it is necessary and insightful to evaluate daylighting of housing from the perspective of carbon emission reduction. In this paper, three design control factors of window height, window/wall ratio and aspect ratio of window are studied. Several preliminary design optimization strategies based on residential lighting in Shanghai are proposed.

1. Introduction

For carbon emission of housing, it is affected by many factors. Among them, as an early phase of decision making, design plays a decisive role affecting the construction phase and use situation coming afterwards. Therefore, carbon emission should be taken into account at the initial phase of construction and also as an important design basis and standard for evaluating.

In the current situation of China, the main problems of daylighting design of urban housing mainly include two aspects: on the one hand, it just stays at the level of pandering to relevant existing building codes; on the other hand, it tends to be more blind pursuit of transparency [1]. Both of the above may cause the actual use of housing to fail to meet the design expectations and bring sensory experience and environmental problems related to energy consumption and carbon emissions [2]. In this paper, energy consumption and daylighting simulation by using design aid software combined with empirical analysis, questionnaire investigation are applied to study the daylighting design related control elements. On this basis, the optimization strategies of daylighting design of housing are proposed to try to achieve a "win-win" of carbon emission reduction and good daylighting in housing projects.

This paper is supported by the "National Natural Science Foundation of China Project: A Study on Sensitivity of Carbon Emission of Residential Building Shape Coefficient through Empirical Researches of Built Housing in Yangtze River Delta (Number.51478315)".

2. Software Simulation

The research object of this paper is a typical middle-rise housing project in Shanghai (figure 1), using it as a reference, the software simulation study is carried out. PKPM-Daylight and PKPM-PBECA, developed and commonly used in contemporary China as design aid tools, are selected as simulation software. The research variables are window sill height, window/wall ratio and aspect ratio of window [3]. After process of modelling and parameter settings, the simulation outcomes of each variable are gathered. The main standards for daylighting condition in this research are average daylighting coefficient and daylighting distribution conditiions [4, 5].

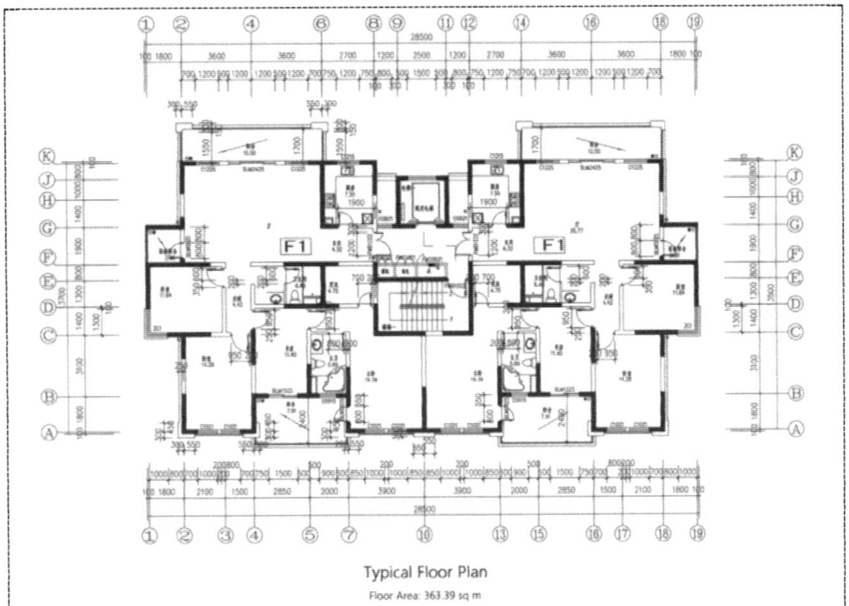

Figure 1. The typical plan of the selected middle-rise housing project.

2.1. *Height of window sill*

The following figures (figure 2-figure 5) show the influence of changing the height of window sill of each orientation on the daylighting coefficient and energy consumption during the use phase. It can be seen that the influence of changing the height of window sill on energy consumption is very small and can be ignored. The daylighting coefficient increases with the increase of the height of the window sill, reaching its peak value and inflection point when the height of window sill is 0.90m.

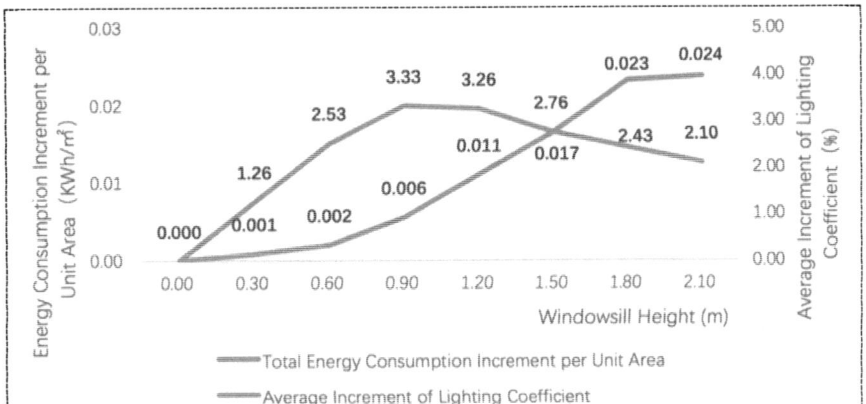

Figure 2. Increments of daylighting coefficient and total energy consumption influenced by east windowsill heights.

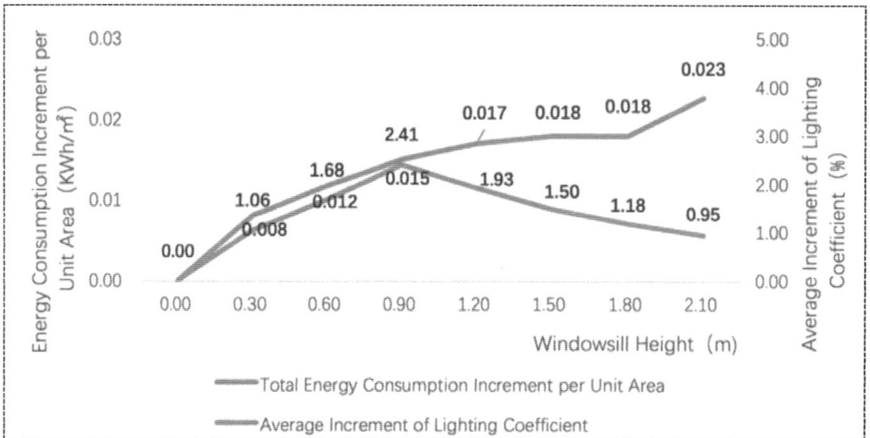

Figure 3. Increments of daylighting coefficient and total energy consumption influenced by south windowsill heights.

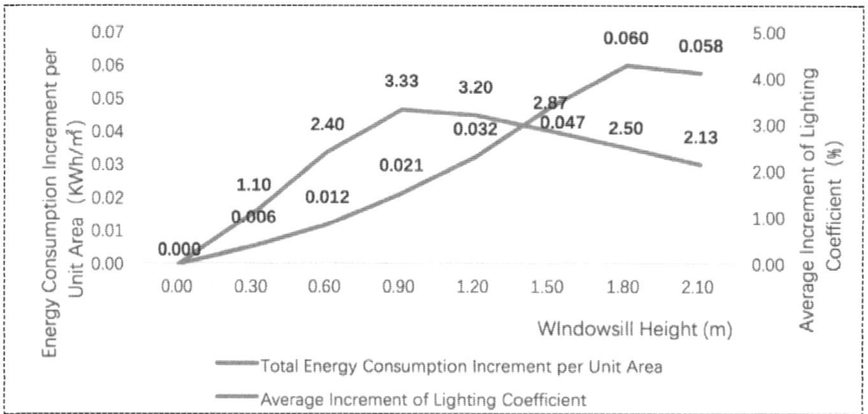

Figure 4. Increments of daylighting coefficient and total energy consumption influenced by west windowsill heights.

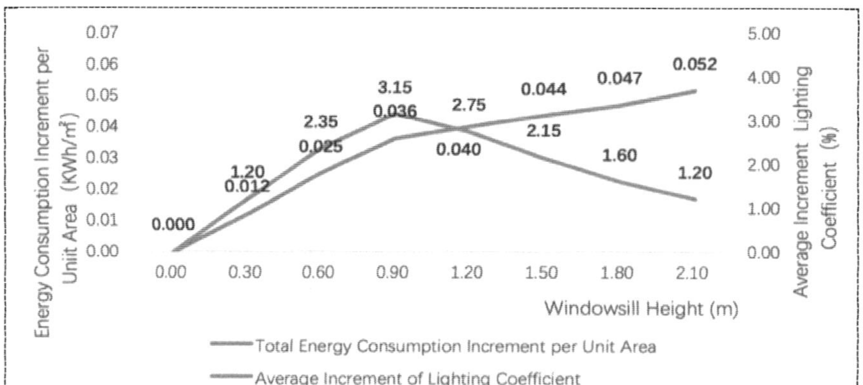

Figure 5. Increments of daylighting coefficient and total energy consumption influenced by north windowsill heights.

2.2. *Window/wall ratio*

By adjusting the window/wall ratio of each orientation, that is, changing the total window area of each orientation will significantly affect the lighting effect and the energy consumption during the use phase of the house, and the increase will basically show a linear trend. In terms of orientation, when the window/wall ratio is increased, the order of the increment of carbon emissions per unit from large to small is south > east ≈ west > north, the order of the increase of indoor average daylighting coefficient is east ≈ west > north > south. Therefore, when the window/wall ratio is considered simply from the view of the lighting design of the house, it is an effective strategy to increase the window wall ratio of each orientation.

2.3. *Aspect ratio of window*

The effect of changing the aspect ratio of window, the ratio of window length to width, on energy consumption and carbon emissions is negligible in the case of the same window/wall ratio. However, from the perspective of indoor daylighting coefficient, when the window/wall ratio is less than 0.30, increasing the window width has a significant advantage over the increase of window height in increase of the lighting coefficient. The difference between these two can be nearly seven times based on the simulation results. When the window/wall ratio is greater than 0.30, the advantages of adjusting window width in increasing daylighting coefficient still exist. It is only that the difference between the two decreases gradually with the increase of window width and window height until it finally disappears (figure 6-figure 13).

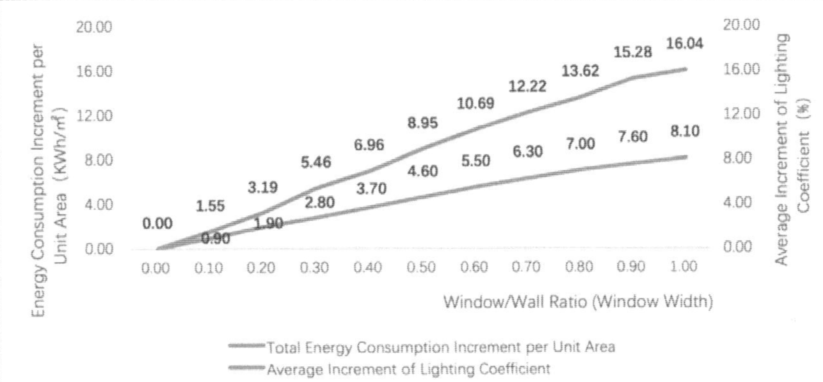

Figure 6. Increments of lighting coefficient and total energy consumption influenced by east window/wall ratio (window width).

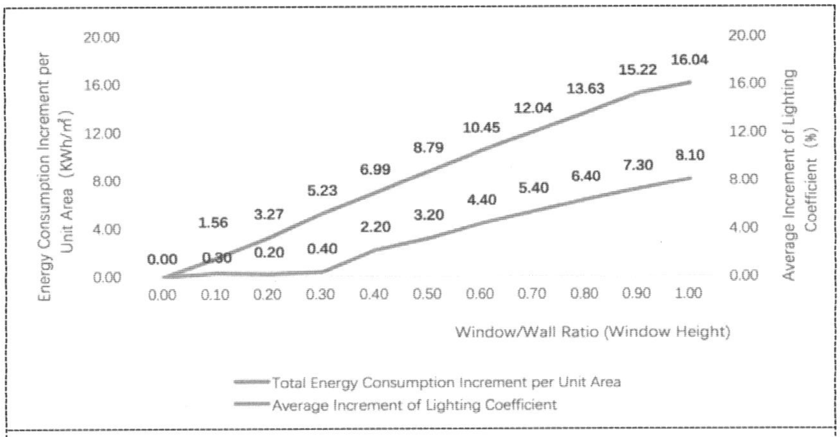

Figure 7. Increments of lighting coefficient and total energy consumption influenced by east window/wall ratio (window height).

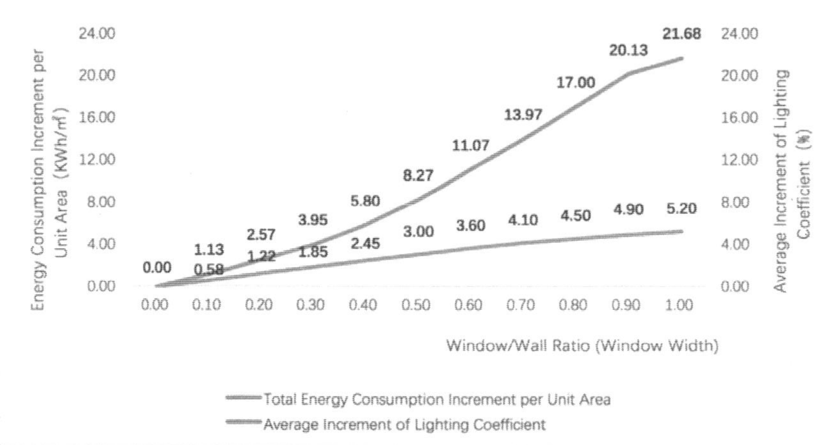

Figure 8. Increments of lighting coefficient and total energy consumption influenced by south window/wall ratio (window width).

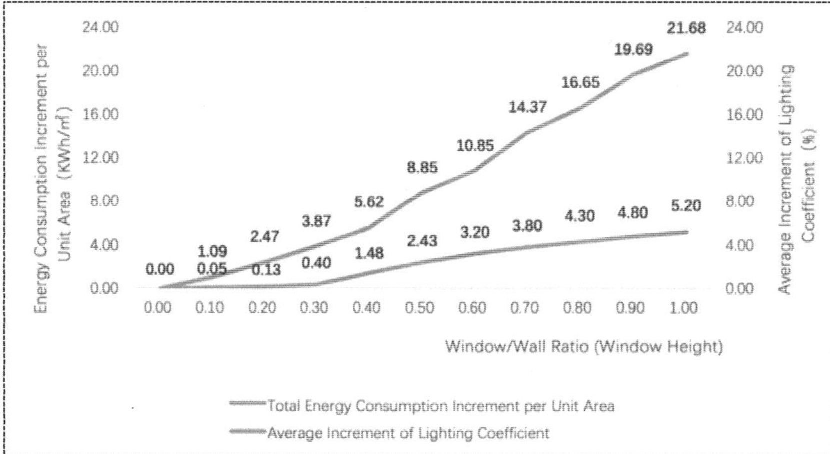

Figure 9. Increments of lighting coefficient and total energy consumption influenced by south window/wall ratio (window height).

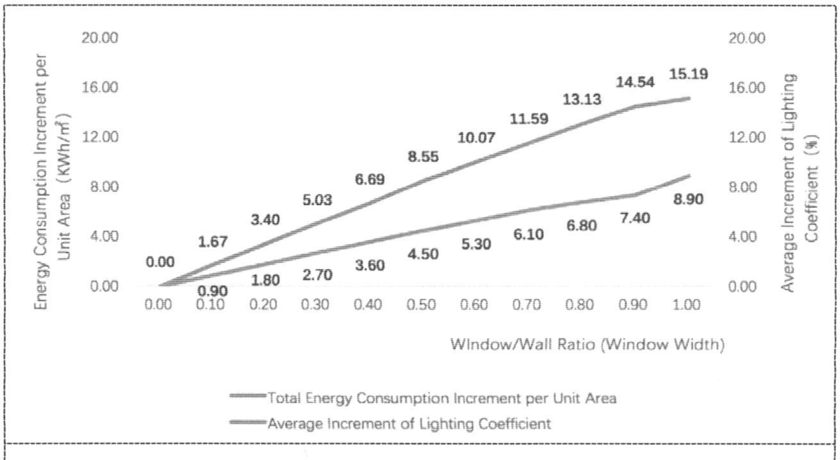

Figure 10. Increments of lighting coefficient and total energy consumption influenced by west window/wall ratio (window width).

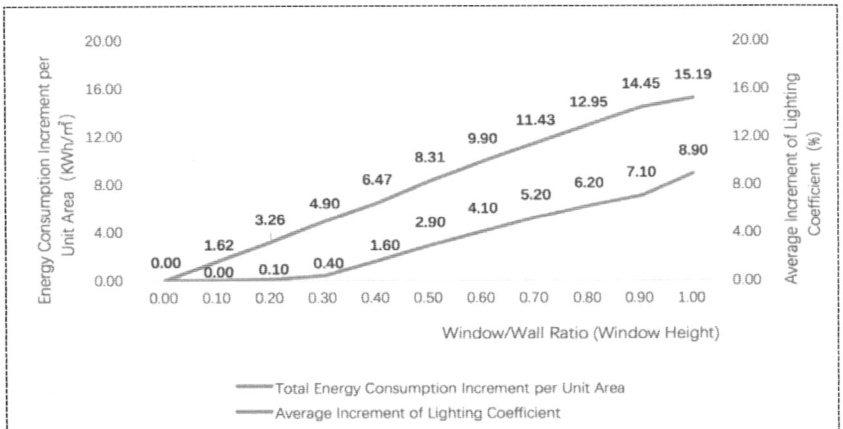

Figure 11. Increments of lighting coefficient and total energy consumption influenced by west window/wall ratio (window height).

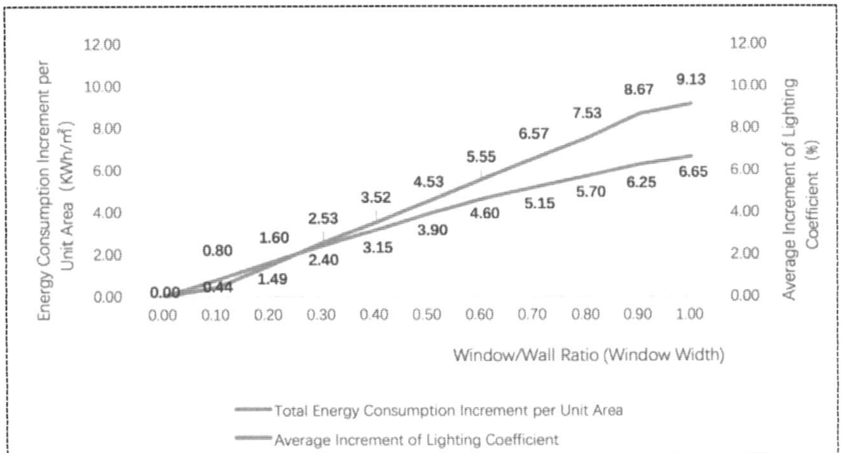

Figure 12. Increments of lighting coefficient and total energy consumption influenced by North window/wall ratio (window width).

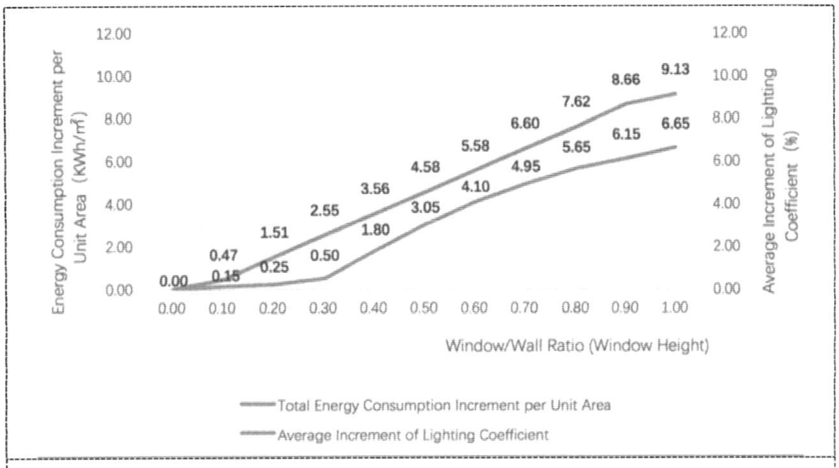

Figure 13. Increments of lighting coefficient and total energy consumption influenced by North window/wall ratio (window height).

3. Calculation of Carbon Emissions

The carbon emission calculation system of this paper is derived from Carbon Emission Calculation Standard CECS 374-2014 in China [6]. On the basis of this specification, the scope of carbon emission units studied in this paper is reduced to the following two aspects which have the highest correlation with residential lighting, and the sum of the two is the number of residential carbon emissions in the whole life cycle (formula 1):

$$\Delta E=\Delta E_M+\Delta E_U \tag{1}$$

In this formula, ΔE is the change of carbon emission ($kgCO_2eq$) in whole life cycle due to residential lighting related variables; ΔE_M is the change of carbon emission ($kgCO_2eq$) caused by relevant lighting design variables of production, construction, demolishing and recycling phase of residential building; ΔE_U is the change of carbon emission ($kgCO_2eq$) in the operation phase of residential building due to residential lighting design variables. Among them, ΔE_M can be calculated by consulting the Athena Eco Calculator for Residential Assemblies database. ΔE_U can be simulated and calculated by PKPM-PBECA energy consumption software.

3.1. Calculation of carbon emissions in use phase

According to the national grid emission factors released by the National Climate Change Agency of the National Development and Reform Commission, the carbon emission can be calculated by bringing the energy consumption into the following formula 2 and formula 3:

$$\Delta E_U=EF_E\times\Delta C_E \tag{2}$$

$$EF_E=(EF_{grid,\ OM,\ y}+EF_{grid,\ BM,\ y})/2 \tag{3}$$

ΔE_U represents the carbon emission change of residential building in use phase ($kgCO_2eq$); EF_E represents the grid baseline emission factor in Shanghai area ($kgCO_2eq/KWh$); ΔC_E represents the annual total energy consumption difference of residential building in use phase (KWh); $EF_{grid,\ OM,\ y}$ represents the electricity marginal emission factor of the regional grid ($kgCO_2eq/KWh$); $EF_{grid,\ BM,\ y}$ represents the capacity marginal emission factor of the regional grid ($kgCO_2eq/KWh$).

Table 1. Regional power grid division of China

Region	Provinces and Cities
East	Shanghai, Jiangsu Province, Zhejiang Province, Anhui Province, Fujian Province

Table 2. Regional grid emission factor of China, 2015

Regional Grid	$EF_{grid,OM,y}$ ($kgCO_2eq/KWh$)	$EF_{grid,BM,y}$ ($kgCO_2eq/KWh$)
East	0.8112	0.5945

3.2. Calculation of carbon emissions in use phase

As can be seen from the tables above (table 1 and table 2), Shanghai belongs to the east regional grid division. According to the data from the tables, the EF_E of east regional grid is 0.7029 ($kgCO_2eq/KWh$). With the energy consumption data taken into the formula 2 respectively, the carbon emission data is calculated. The window/wall ratio data can be considered as an average value of the counterpart of width and height. The outcomes are listed in the following tables (table 3 and table 4).

Table 3. Annual increment of carbon emissions influenced by windowsill height in use phase

Windowsill Height (m)	0.00	0.30	0.60	0.90	1.20	1.50	1.80	2.10
East ($kgCO_2eq/m^2$)	0.0000	0.0006	0.0014	0.0039	0.0078	0.0116	0.0163	0.0167
South ($kgCO_2eq/m^2$)	0.0000	0.0057	0.0084	0.0106	0.0120	0.0127	0.0127	0.0159
West ($kgCO_2eq/m^2$)	0.0000	0.0039	0.0082	0.0149	0.0227	0.0331	0.0421	0.0404
North ($kgCO_2eq/m^2$)	0.0000	0.0086	0.0176	0.0255	0.0282	0.0306	0.0331	0.0363

Table 4. Annual increment of carbon emissions caused by window/wall ratio in use phase

Window/Wall Ratio	0.00	0.10	0.20	0.30	0.40	0.50	0.60	0.70	0.80	0.90	1.00
East ($kgCO_2eq/m^2$)	0.00	1.10	2.27	3.76	4.90	6.24	7.43	8.53	9.58	10.72	11.27
South ($kgCO_2eq/m^2$)	0.00	0.78	1.77	2.75	4.02	6.02	7.71	9.96	11.83	14.00	15.24
West ($kgCO_2eq/m^2$)	0.00	1.16	2.34	3.49	4.63	5.93	7.02	8.09	9.17	10.19	10.68
North ($kgCO_2eq/m^2$)	0.00	0.32	1.06	1.79	2.48	3.20	3.91	4.63	5.32	6.09	6.42

3.3. The calculation of carbon emission in the phase of materialization and demolition

Each residential building needs to go through a long process of materialization, from production, manufacturing, processing and transportation of different materials and components; assembling and construction; at last, the demolition phase, which includes taking care of the remaining, recycling and so on. All of these processes are accompanied by the carbon emissions. This section will focus on calculating the carbon emission changes in the materialization and demolition phases of residential buildings caused by lighting design.

Residential buildings, like most other types of buildings, consist of envelop structure and supporting structure. In previous software simulations, windowsill height and window/wall ratios were adjusted based on the reference model with other parameters staying unchanged, thus, it can be regarded as no change in the supporting structure. Therefore, the carbon emissions in the materialization and demolition phases are reflected in the changes of the envelope structure.

It can be considered that changing the windowsill height does not have influence on material change. While changing the window/wall ratio results in a change of the window area and wall area due to the total area is a fixed constant for a building. Due to the fact that the materialization and demolition processes of different envelope structures are also different, the changes of the amount of different envelope structures are the main factors that change the carbon emissions of residential buildings in these phases. The envelope materials of the reference residential building are concrete block and insulated aluminium alloy window frame with double low-e hollow glazing. Thus, the carbon emission caused by window/wall ratio can be calculated by applying the formula 4, 5 listed below:

$$\Delta E_{MG}=\Delta R_O \times S_O \times C_{MC} \tag{4}$$

$$\Delta E_{MC}=(1-\Delta R_O) \times S_O \times C_{MC} \tag{5}$$

ΔR_O represents the window-wall ratio; S_O represents the envelope area of each orientation (m^2); ΔC_{MG} is the reference value of carbon emission per unit area ($kgCO_2eq/m^2$) per unit area of insulated aluminium alloy window frame with double low-e hollow glazing; ΔC_{MC} is the carbon emission per unit area of concrete block wall ($kgCO_2eq/m^2$).

The reference values for the materials used as envelope structure are obtained from the database which is widely used in the United States, the Athena Eco Calculator for Residential Assemblies, as mentioned before. This database collects a large number of homes in the United States, after conducting a huge amount of calculations and material statistics, a systematic database for carbon emission calculation is completed. The following calculation of the materialization and demolition phase will be conducted on a basis of this database.

As mentioned before, the adjustment of the windowsill height does not affect the amount of material used for the building envelope. Thus, it can be considered that different settings of windowsill height do not cause any carbon emission changes in materialization and demolition phase.

D. EXTERIOR WALLS

IN THE YELLOW CELLS BELOW, ENTER THE AMOUNT OF SQUARE FOOTAGE THAT EACH ASSEMBLY USES IN YOUR BUILDING

	WALL TYPE	WALL ENVELOPE	Square footage	Percentage of total	Fossil Fuel Consumption per ft² (MJ)	Global Warming Potential per ft2 (kg CO2 eq)	Acidification Potential per ft² (moles of H+ eq)
Average across exterior wall assemblies:					**121.40**	**8.60**	**3.28**
8" CONCRETE BLOCK					**121.05**	**9.83**	**2.90**
1	**Concrete Block**	Clay Brick Cladding w/ 1" Air Space R5 XPS Continuous Insulation 1/2" Gypsum Board + 2 Coats Latex Paint	0.0		120.02	10.09	3.40
2	**Concrete Block**	Metal Cladding R5 XPS Continuous Insulation 1/2" Gypsum Board + 2 Coats Latex Paint	0.0		194.04	15.86	5.03
3	**Concrete Block**	2 Coat Stucco Over Porous Surface R5 XPS Continuous Insulation 1/2" Gypsum Board + 2 Coats Latex Paint	0.0		79.48	7.07	1.83
4	**Concrete Block**	Vinyl Siding R5 XPS Continuous Insulation 1/2" Gypsum Board + 2 Coats Latex Paint	0.0		113.49	8.35	2.68

Figure 14. Carbon emission factor of wall per unit area in the Athena database.

E. WINDOWS

IN THE YELLOW CELLS BELOW, ENTER THE AMOUNT OF SQUARE FOOTAGE THAT EACH ASSEMBLY USES IN YOUR BUILDING

	FRAME TYPE	DOUBLE GLAZING TYPE	Square footage	Percentage of total	Fossil Fuel Consumption per ft² (MJ)	Global Warming Potential per ft2 (kg CO2 eq)	Acidification Potential per ft² (moles of H+ eq)
Average across all window types:					**514.92**	**45.33**	**31.16**
1	Aluminum - Operable	Low E, Argon Filled	0.0		798.29	67.63	56.56
2	Vinyl-clad Wood - Operable	Low E, Argon Filled	0.0		371.31	34.33	21.16
3	Vinyl - Operable	Low E, Argon Filled	0.0		519.25	43.69	24.85

Figure 15. Carbon emission factor of window per unit area in the Athena database.

As shown in the figures listed above (figure 14 and figure 15), it can be seen that the closet wall material is the No.3 type wall (which is concrete block, 2 coat stuccos over porous surface, R5 XPS continuous insulation). Its GWP value is $7.07kgCO_2eq/ft^2$, which converted to a metric unit is $78.56kgCO_2eq/m^2$; similarly, the GWP of insulated aluminium alloy window frame with double low-e hollow glazing (aluminium operable low-e double glazing) is $67.63kgCO_2eq/ft^2$, which converted to metric units is $751.44kgCO_2eq/m^2$.

$$\Delta E_M=(\Delta E_{MG}+\Delta E_{MC})/S_C \tag{6}$$

ΔE_M represents the change of carbon emission per unit area ($kgCO_2eq/m^2$) caused by envelope structure change in materialization and demolition phase; ΔE_{MG} represents the change of carbon emission caused by the area change of aluminium operable low-e double glazing ($kgCO_2eq$); ΔE_{MC} represents the change in carbon emissions caused by concrete block walls with XPS insulation

($kgCO_2eq$); S_C represents the total floor area (m^2). Taking the related parameters into formula 4, 5 and 6 respectively for calculation, the outcomes are listed below (table 5).

Table 5. Increment of carbon emissions influenced by window/wall ratio in materialization and demolition phase

Total Floor Area 3442.17m^2				East/West Surface Area 799.22m^2				South/North Surface Area 967.98m^2			
Window/Wall Ratio	0.00	0.10	0.20	0.30	0.40	0.50	0.60	0.70	0.80	0.90	1.00
East ($kgCO_2eq/m^2$)	0.00	15.62	31.25	46.87	62.49	78.12	93.74	109.36	124.99	140.61	156.23
South ($kgCO_2eq/m^2$)	0.00	18.92	37.85	56.77	75.69	94.61	113.54	132.46	151.38	170.30	189.22
West ($kgCO_2eq/m^2$)	0.00	15.62	31.25	46.87	62.49	78.12	93.74	109.36	124.99	140.61	156.23
North ($kgCO_2eq/m^2$)	0.00	18.92	37.85	56.77	75.69	94.61	113.54	132.46	151.38	170.30	189.22

3.4. The calculation of carbon emission in the phase of materialization and demolition

According to the formula 1 proposed previously, the carbon emission changes in whole life cycle can be calculated by taking all results gathered above in materialization phase, use phase, and demolition phase. To simplify the process, the use phase is calculated as 50 years. The final results are listed as below (table 6 and table 7). As can be seen from the chart above, the carbon emission increment caused by window/wall ratio are significantly more than the counterpart of windowsill. Therefore, the window/wall ratio may be the most potential aspect that should be pay more attention to when dealing with carbon emission issues of residential buildings.

Table 6. Increment of carbon emissions influenced by windowsill height in whole life cycle (50 years)

Windowsill Height（m）	0.00	0.30	0.60	0.90	1.20	1.50	1.80	2.10
East ($kgCO_2eq/m^2$)	0.00	0.03	0.07	0.19	0.39	0.58	0.82	0.84
South ($kgCO_2eq/m^2$)	0.00	0.29	0.42	0.53	0.60	0.63	0.63	0.80
West ($kgCO_2eq/m^2$)	0.00	0.19	0.41	0.75	1.13	1.65	2.10	2.02
North ($kgCO_2eq/m^2$)	0.00	0.43	0.88	1.28	1.41	1.53	1.65	1.82

Table 7. Increment of carbon emissions influenced by window/wall ratio in whole life cycle (50 years)

Window/Wall Ratio	0.00	0.10	0.20	0.30	0.40	0.50	0.60	0.70	0.80	0.90	1.00
East ($kgCO_2eq/m^2$)	0.00	70.62	144.75	234.87	307.49	390.12	465.24	535.86	603.99	676.61	719.73
South ($kgCO_2eq/m^2$)	0.00	57.92	126.35	194.27	276.69	395.61	499.04	630.46	742.88	870.30	951.22
West ($kgCO_2eq/m^2$)	0.00	73.62	148.25	221.37	293.99	374.62	444.74	513.86	583.49	650.11	690.23
North ($kgCO_2eq/m^2$)	0.00	34.92	90.85	146.27	199.69	254.61	309.04	363.96	417.38	474.80	510.22

4. Carbon emission reduction potential of household behaviour influenced by daylighting design

4.1. Questionnaire Survey

The previous parts of this paper define and discuss the main influencing factors of residential carbon emissions in this research, which can be classified as the materialization and the demolishment phase and the use phase. Among them, the carbon emission of the use phase is not only related to the energy

consumption of the equipment which is determined by the physical performance of the residential enclosure structures, but also closely related to the household behaviour, like habits of using household appliance and other equipment [7]. The impact of household behaviour on residential carbon emissions is often studied isolated or ignored, systematic, comprehensive demonstration and research need to be further explored. With the research perspectives being continuous widening in recent decades, more and more scholars continue to raise awareness of energy conservation and carbon emission reduction in many areas. With sociology, behaviour, environmental psychology and other multidisciplinary fields and other more perspectives emerging, the majority of scholars are gradually study in this area deeper. Household behaviour has a potential impact on residential carbon emissions in residential buildings, which makes it worth to be explored [7, 8].
The above section outlines the impact of household behaviour on residential carbon emissions. In order to explore further, a questionnaire survey is conducted, as a supplementary study for simulation, to try to understand how residential lighting, household behaviour and emission reduction affect each other in hot summer and cold winter area [9,10]. The entire research system is enriched by analysing and summarizing the survey results, screening out the key factors.

4.2. Analysis of outcome

As for why using artificial lighting, only 11.1% of the sample preferred artificial lighting [11]. More people are forced to use artificial lighting because of the inadequate daylighting and poor lighting quality. The factors of residential design will also have a certain impact on the use of equipment. The shading, for instance, is a common method for controlling direct sunlight and radiant heat into the interior in summer times, the effect of the physical characteristics of residential buildings is without doubt. As can be seen from figure 16 and figure 17, the potential behaviour impact on household has been illustrated. In units with shading, the family members have a higher preference for passive cooling ways to enjoy a comfortable natural climate than the members living in the houses without any shading equipment. Equipped with the exterior louvers, household behaviour of air conditioning in the summer has been shown an increment proportion to open intermittently than the counterparts of the units without shading (11.11%); while the proportion of household to set a low running temperature of cooling equipment decrease significantly as well; at the same time, the proportion of household to set high temperature of air conditioning in summer is relatively high in shading residential, especially for exterior shading louvers (35.29%). On one hand, among all forms of shading types shown in the figure above, the exterior shading types such as louvers, balcony and building shape shading have more influence in changing the behaviour of setting the operating temperature of cooling equipment in summer than the interior shading form. On the other hand, conversely, when heating equipment was used in winter, the proportion of household to set operating temperatures in residential units without shading show a higher number in operating their devices intermittently (22.22%) than the counterpart of the residential units with shading.

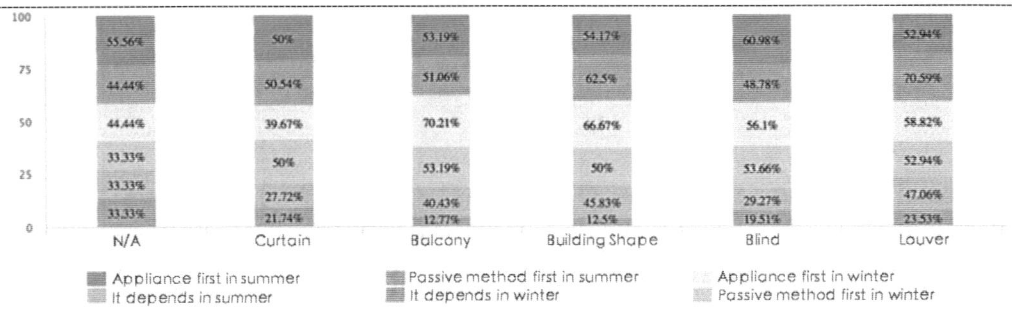

Figure 16. Shading and cooling, heating preference.

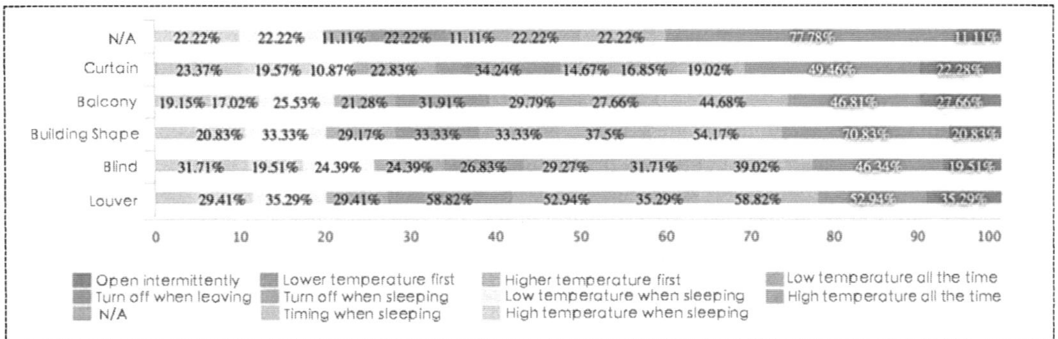

Figure 17. Shading and behaviour of cooling equipment.

5. Summary

The research work of this paper is conducted by applying simulation, empirical study, field investigation and research and so on. Through the quantitative analysis, the possibility of carbon emission reduction in residential lighting design is discussed in a certain framework. The main conclusions are summarized as follows.

(1) Height of windowsill. Based on the simulation and empirical research, when the window area is a fixed value, the most cost-effective approach to improve the indoor lighting coefficient is to increase the height of the windowsill appropriately, and the effect gets to the best when height of the windowsill reaches about 0.90m; after this turning point, the lighting coefficient goes down with height of the windowsill increasing; the evenness of indoor daylighting distribution improved with the height of the windowsill increasing; as for the carbon emission consideration, with the increment of indoor lighting coefficient, the carbon emission increment is so small that can be ignored. Combining these two aspects, design an appropriate windowsill height can be an efficient way in residential design for improving indoor lighting conditions to some extent without causing significant carbon emission increment, which is a suitable sustainable design strategy under the context of carbon emission reduction. And the recommendation height for windowsill is 0.90m.

(2) Window/wall ratio. By adjusting the window/wall ratio of different orientations, that is, changing the window area in each direction respectively when the total area of each direction is a fixed value, the residential lighting efficient will be significantly affected. Increasing the window/wall ratio is a direct way to improve the residential lighting condition, but it is also accompanied by a considerable increment in carbon emissions in both use, materialization and demolition phase. And the growth rate between the carbon emission increment and lighting coefficient shows a linear correlation; taking the direction of the window into consideration, the order of carbon emission increment per unit is south > east ≈ west > north; the order of average daylighting coefficient increment is east ≈ west > north > south. The recommendation window/wall ratio for south direction can be summarized as 0.40-0.50.

(3) Aspect ratio of window. When the window/wall ratio is a fixed value, or in other words, the total area of window is a fixed value, the carbon emission increment caused by changing the shape of window in each direction is almost negligible. However, the shape of window does make a notable difference when it comes to the indoor lighting coefficient distribution. When window/wall ratio is less than 0.3, increasing the window width has significant advantages for increasing the lighting coefficient compared with increasing the window height, the difference between these two is almost up to 7 times; while window/wall ratio is more than 0.3, the advantages of adjusting the window width in increasing the indoor natural daylighting effect still exist, but the differences between these two gradually decrease. Given the carbon emission can be ignored for adjusting the shape of the window, which in most situations, widths and heights in residential buildings, if other design conditions permit, the better design strategy of improving lighting from the carbon emission perspective is to increase the window width first instead of increasing the window height. Especially when the window height or the height of windowsill is below 30% or less of the floor height.

In addition, through a series of qualitative analysis of questionnaire survey, it can be concluded that residential lighting design has the potential to reduce carbon emissions by influence the household behaviour toward the operation of cooling and heating equipment, such as proper shading design will affect the cooling habits and reduce household reliance on cooling equipment so that it will reduce carbon emission in use phase to some extent.

In summary, the daylighting design in the context of carbon emission reduction is a relatively complicated and multi-variable comprehensive consideration. And a considerate and appropriate lighting design has many potentials in balancing carbon emissions and lighting conditions.

6. References

[1] Eric S, Jia H and Maulin P 2014 energy and visual comfort analysis of lighting and daylight control strategies *Building and Environment* **78** 155-70

[2] Ge C and Xiong D 2009 study on the design of daylighting in modern residential buildings *House Science and Technology* **03** 24-7

[3] Stevenson F, Isabel C, Hancock M 2013 the usability of control interfaces in low carbon housing *Architecture Science Review* **56** 70-81

[4] Zhang B, Li G and Zhao J 2010 study on the design of residential lighting based on the concept of lighting efficiency *Journal of Lighting Engineering* **04** 14-8

[5] Popoolaa O, Mundaa J and Mpandaa 2015 a residential lighting load profile modelling *Energy and Buildings* **90** 29-40

[6] Zhang L, Huang Y and Huang X 2012 construction life cycle carbon evaluation based on standard computing platform *Huazhong Architecture.* **06** 32-4

[7] Yujiro H, Tomohiko I and Yukiko Y 2016 estimating residential CO2 emissions based on daily activities and consideration of methods to reduce emissions *Building and Environment* **103** 1-8

[8] Shan S and Bin Z 2016 occupants' interactions with windows in 8 residential apartments in Beijing and Nanjing, China *Building Simulation* **9** 221-31

[9] Peng X, Mak C and Cheung D 2014 the effects of daylighting and human behavior on luminous comfort in residential buildings: a questionnaire survey *Building and Environment* **81** 51-9

[10] Valentina F, Rune A and Stefano P 2013 a methodology for modelling energy-related human behavior: application to window opening behavior in residential buildings *Building Simulation* **6** 415-27

[11] Acosta I, Campano M and Molina J 2016 window design in architecture: analysis of energy savings for lighting and visual comfort in residential spaces *Applied Energy* **168** 493-506

Acknowledgments

I would like to express my gratitude to my beloved supervisor-Yiru Huang for this opportunity to be part of this significant research program and Clark E Llewellyn for his guidance.

IOP Conf. Series: Earth and Environmental Science **182** (2018) 012018 doi:10.1088/1755-1315/182/1/012018

Research on Evaluation and Factors of Regional Green Innovation Performance Based on ER-XIANG Dual Theory

Chaojun Yang[1], Wenke Yang[1], Ruoqing Hu[1]

1 Faculty of Management and Economics, Kunming University of Science and Technology, Kunming, CHN
E-mail: wenkedyang@foxmail.com

Abstract. On the basis of panel data from 30 provinces during 2005-2014, the index system of regional green innovation performance is built from efficiency and output two dimensions based on dual theory. The evaluation of regional green innovation performance is using PPE-Malmquist-LWN model synthetically and the empirical analysis is used to study on factors of regional green innovation performance. The results indicate that regional green innovation performance shows an M type trend, and in east is higher than that in west. On the whole, FDI, ODI, environmental regulation of lag 1 have positive effects on regional green innovation performance, but internal and externa 1 market needs, intellectual property protection, market institutions have negative effect. In eastern region, FDI, external market needs, market institutions have negative effect, but environmental regulation of lag 1 has positive effects. In central region, internal market needs has negative effect, but ODI and environmental regulation of lag 1 have positive effects. In western region, internal market needs and intellectual property protection have negative effect, but FDI, ODI and external market needs have positive effects.

1. Introduction

With the continuous development of economic globalization and the increasing problem of resource and environment, green innovation has become the decisive force for the sustainable development of the global economy. The Chinese government emphasizes the coordinated development of economic and social and eco-environmental systems by the implementation of green innovation nowadays. To China's vast territory, regional resources endowment differences affect the balanced development of regional economy and the regional coordinated development of resources and environment. Therefore, the accurate evaluation of regional green innovation performance will help understand the green innovation performance gap with similar regions, and clearly enhance the influencing factors of green innovation performance and promote the sustainable development of China.

At present, domestic and foreign scholars have obtained more research achievements in the research of green innovation performance evaluation and its influencing factors. In terms of the performance evaluation of green innovation, most of the literatures are from the perspective of output of green innovation performance evaluation index system, and then using the factor analysis model(Cheng and Liao, 2011) [1], projection pursuit model(Bi et al., 2013) [2], the Super - SBM model(Wang et al., 2016) [3]to comprehensive evaluation method, the main difference lies in the different index system.

The index system constructed includes three dimensions: ecological organization, ecological technological innovation and eco-product innovation (Cheng and Shiu, 2012) [4]. However, green innovation performance evaluation index system constructed by management innovation, technological innovation, product innovation and technological innovation(Tseng et al., 2013) [5].

Some literatures use DEA and its derivative methods to calculate the input-output efficiency of green innovation from the perspective of efficiency. (Wang et al., 2016) [3].

In the study of the influence factors of green innovation, environmental regulation is considered to be one of the most important influence factors, but the effects of environmental regulation on green innovation performance remains controversial. Researchers finds that environmental regulation promotes green innovation performance by changing the green behaviour of enterprises (Huang et al., 2016) [6]. And the research results of a group shows that the pollutant discharge permit system and the time limit governance system in the environmental system have negative impacts (Fan et al., 2013) [7]. However, A typical research shows that there is a u-shaped relationship between environmental regulation and innovation(Shen et al., 2012) [8]. In addition, the market needs(Zailani et al., 2015) [9], technological advancements (Fan et al., 2013) [7], environmental management systems(Cuerva et al., 2013) [10], customer stress and corporate green response behaviours(Huang et al., 2016) [6]are also thought to be a key factor in promoting the performance of green innovation.

To sum up, the existing literature has achieved many findings on green innovation performance evaluation and influencing factors, but we believe that it could be improved in the following aspects: firstly, the evaluation result based on the single perspective of output or efficiency may lead to the false positive judgment of green innovation performance; Secondly, the green innovation performance influence factors are relatively random, lacking of comparison in the influence factors of green innovation performance among different regions.

Therefore, we use 30 provinces panel data during 2005-2014, based on the duality characteristics of the perspective of output and efficiency to build green innovation performance evaluation index system, using the PPE - Malmquist - LWM model for comprehensive evaluation; Then, the influence factors of green innovation performance were extracted from the relatively intact framework, and the differences of the influence factors were analysed from the whole, eastern, central and western regions. It is of great practical significance to accurately grasp the status of green innovation performance and its influence factors in various provinces of China, and to formulate the policy of green innovation.

2. Research on the Performance Evaluation of Regional Green Innovation Based on ER-XIANG Dual Theory

2.1. Regional green innovation performance evaluation index system based on ER-XIANG duality

Based on the wave particle duality and system theory, we argue that green innovation performance can be divided into two dimensions: green innovation output and green innovation efficiency, namely, the two image characteristics of green innovation performance. In this case, the green innovation is the actual dimension of green innovation, which are the socioeconomic performance and the environmental benefit including the output of green patents, the technical skills, the knowledge of the products, the value of the new product sales, and the output of the green innovation. Green innovative output represents the quantitative result of green innovative performance from the perspective of "quantity", and it is a static evaluation on green innovation performance. Green innovation efficiency is a virtual image of green innovation performance, reflecting the effectiveness of the process management of green innovation activities. It is the ability of green innovation subjects to integrate and coordinate various green innovation resources into new products, new technology and new services in the green innovation activities. In the evaluation, green innovation efficiency is reflected in the input and output relationship of green innovation activities, reflecting the process management level of technological innovation performance from the perspective of "quality". In a word, the higher the green innovation performance is not only characterized by high green innovation output, also show the good efficiency of green innovation. According to the above analysis, the green innovation performance evaluation indicator system based on ER-XIANG Dual theory includes two first-grade indexes: green innovation output and green innovation efficiency. Among them, secondary indicators of green innovation output includes green patent application quantity, new product sales revenue, technology market turnover, pollutant discharge fee of unit industrial output value, energy consumption in unit area output value, industrial waste removal rate, industrial waste generation intensity and industrial waste recycling rate; The green innovation efficiency is calculated by the input and output index, and the secondary indicators of green innovation inputs include the R&D capital investment, the new product development costs, the environmental pollution management investment, the technology import and the technology modification costs, full time equivalent of R&D personnel and the number of stuff in the environmental protection system at the end of the year.

2.2. Evaluation model based on PPE-Malmquist-LWM

The PPE-Malmquist-LWM model was used to evaluate green innovation output, green innovation efficiency and green innovation performance. Projection tracing (PPE) is an exploratory data analysis method driven directly by sample data and is particularly suitable for analysing and processing non-linear and non-normal high dimensional data. The projection tracking evaluation model is used for comprehensive evaluation. Calculating the projection pursuit direction is a complicated nonlinear optimization problem, so it is difficult to be solved by the traditional optimization method. Therefore, we use the real-coded accelerating genetic algorithm (RAGA) to find the best projection direction. Based on the existing literatures, we also select the Malmquist index method to measure the efficiency of green innovation from the perspective of the effectiveness of green innovation input and output. The Malmquist index method is a dynamic comparison way based on the relative validity of the DEA model. It uses the ratio of distance function to calculate the input-output efficiency, which reflects the change of input-output efficiency from t period to t+l period.

2.3. Regional green innovation performance evaluation and result analysis.

Based on the above theory, this paper makes an empirical assessment on green innovation performance of our 30 regions in 2005-2014. From the time dimension, we can see the overall green innovation performance evaluation average represented M type volatility growth during 2005-2014 in figure 1.

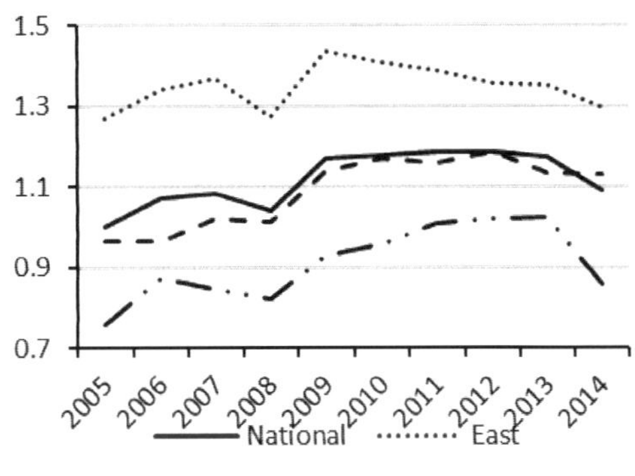

Figure 1. Regional green innovation performance

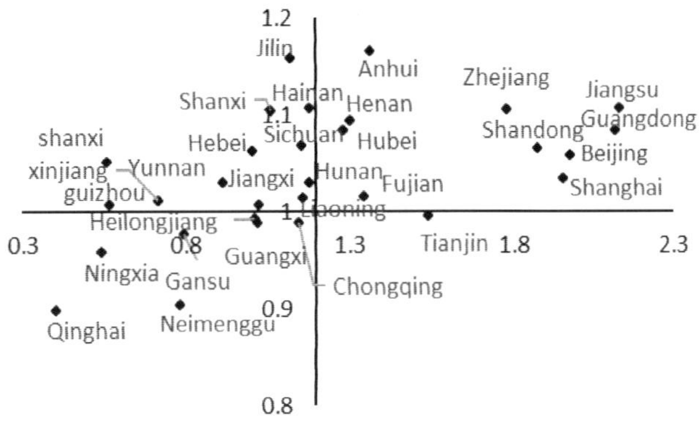

Figure 2. Regional green innovation performance spatial distribution based on ER-XIANG Dual Theory.

From the perspective of the dual characteristics of regional green innovation performance (figure 2, the horizontal axis output =1.198, vertical axis efficiency =1), the green innovation efficiency of 73.33% in the country is in the progressive state (i.e., the green innovation efficiency rating is greater than 1), and 36.67% of the province's green innovation output is larger than the national average.

3. Study on the Influence Factors of Regional Green Innovation Performance

3.1. Model specification, variables and data

Successful green innovation needs to balance the various driven factors on the innovation chain. This paper analyses the influence factors of regional green innovation performance from three aspects: technology, market and system. The measurement model as follows:

$$\begin{aligned}\ln GIP_{it} = & C + \alpha_1 \ln FDI_{it} + \alpha_2 \ln ODI_{it} + \beta_1 \ln IMD_{it} + \beta_2 \ln EMD_{it} \\ & + \theta_1 \ln IPS_{it} + \theta_2 \ln MS_{it} + \theta_3 \ln ER_{it} + \theta_4 \ln ER(1)_{it} + \varepsilon_{it}\end{aligned} \tag{1}$$

In this case, the green innovation performance (GIP) is measured above. The FDI and ODI are the technical factor that affects the regional green innovation, while the IMD and EMD are the market factor, and the IPS, MS and ER are the institutional factors. α_i, β_i, θ_i are the coefficient of regression for each factor, C represents the intercept term, ε_{it} indicates the error term of the whole regression equation, and the subscript *i* and t respectively indicate the selected region and year.

Technical factors. The innovation capacity of a country or region is mainly due to internal independent innovation and international technology spillover. Regional green innovation performance, therefore, tend to be internal research and development activities and the influence of external international technology spillovers, but in view of the above evaluation in the process of considering the role of the independent innovation factors such as R&D, this part only consider foreign direct investment, foreign direct investment and the influence of exogenous technological factors. Among them, foreign direct investment (FDI) is measured by its investment(Yang, 2015) [11], which reflects the spillover effect of foreign direct investment in the region. Outward direct investment (ODI) is measured by the foreign direct investment stock inspection, and test the backward spillover effect it obtained.

Market factors. Market demand is the root cause of innovation activities and the final destination of innovation results. Besides, the market demand of green innovation includes domestic market demand and foreign market demand. Thus, we use the average earnings of urban residents to measure our domestic market demand, and to test the impact of regional earnings on the performance of green innovation. Instead, that external market demand (EDM) use industrial export delivery values as a measure of the proportion of industrial output value, reflecting the impact of local export guidance on green innovative performance(Bi et al., 2013) [2].

Institutional factors. Green innovation can produce positive spillovers in both environmental and economic dimensions, leading to a stronger institutional constraint for green innovation activities. The environment system is regarded as the most important influence factors, this paper adopts the "three simultaneity" system of environmental protection investment measure environment (ER) (Fan et al., 2013) [7]; ER(1) variable represents a year lagged environment system, testing whether there exists hysteresis influence in the system of environmental on the performance of green innovation(Li et al., 2014) [12]. In addition, the market system (MS) and the intellectual property protection system (IPS) also have an important influence on the green innovation performance, which is measured by the proportion of assets of industrial state-owned enterprises and the cumulative percentage of patent infringement cases(Yang, 2014) [13], so as to examine the impact of the marketization level of the region and the protection of intellectual property rights on green innovation performance.

Data from 2005-2014 China statistical yearbook of science and technology, the industrial enterprise science and technology activity statistics, China industrial economic statistical yearbook, China statistical yearbook, the China environment yearbook and the state intellectual property office. The green patent data comes from the state intellectual property office, which can be searched by setting the keywords of green patent, and further screening and identification by looking at the patent summary(Li, 2015) [14]. Tibet was excluded due to incomplete data. Thus, we analyse of the provinces, municipalities and autonomous regions, a total of 30.

3.2. Results and findings

Based on the results of the F and the Hausman test, we used the fixed effect model to analyse the green innovation performance factor in the whole sample, the eastern region, and the central region, and we did some empirical analysis of the western region, using the hybrid effect model, and the results were shown as shown in table 1.

Table 1. The table shows the results of regional green innovation performance factors.

Variable	National	East	Central	West
FDI	0.036**	-0.066**	0.076	0.059***
	(2.327)	(-2.265)	(1.304)	(4.519)
ODI	0.021***	0.007	0.064*	0.045***
	(2.973)	(0.569)	(1.948)	(3.628)
IMD	-0.130***	-0.092	-0.326*	-0.120***
	(-3.598)	(-1.423)	(-1.699)	(-5.832)
EMD	-0.044*	-0.129*	-0.058	0.042**
	(-1.939)	(-1.762)	(-0.910)	(2.078)
IPS	-0.230**	0.101	0.153	-0.316***
	(-2.457)	(0.497)	(0.723)	(-2.768)
MS	-0.159**	-0.379***	-0.061	-0.036
	(-2.221)	(-2.665)	(-0.327)	(-0.300)
ER	-0.011	-0.018	-0.033	-0.006
	(-1.246)	(-1.310)	(-0.952)	(-0.235)
ER(1)	0.038***	0.025*	0.08*	0.016
	(4.059)	(1.743)	(1.936)	(0.575)
R^2	0.892	0.830	0.703	0.565
F-test	10.666***	17.473***	6.537***	1.367
H-test	16.006**	22.681***	-	11.710
Model	Fixed	Fixed	Fixed	Hybrid effect model

Dependent variable: GI.
All explanatory variables are measured pre-treatment.
* $p < 0.10$, ** $p < 0.05$, *** $p < 0.01$

Analysis of regression results based on national samples. In view of that regression result, R^2 of the whole sample is 0.892, and the fit effect is good. Firstly, in terms of technical factors, the regression coefficient of foreign direct investment (FDI) and outward direct investment (ODI) is positive, which means that international technology spillover is obtained through FDI and foreign direct investment in China's regional green innovation activities. On the one hand, as the vehicle of finance and technology, the entry of FDI has increased the investment of green innovation in our region, and has helped green innovation in human resources; At the same time, some foreign enterprises use more environment-friendly production technology and pollution treatment technology, which plays a good demonstration role for China's green innovation, and thus has a positive spillover effect on regional green innovation performance. On the other hand, foreign direct investment (especially foreign direct investment with the motivation of acquiring the advanced technology of R&D intensive host country) can form reverse technology spillover through technology cluster absorption mechanism, overseas market competition mechanism, R&D resource sharing mechanism and technology two-way communication mechanism, thus promoting the promotion of green innovation performance. Secondly, in terms of market factors, the regression coefficient of internal market demand (IDM) and external market demand (EDM) is significantly negative, which means that both urban residents' income level and industrial export are negatively correlated with green innovation performance. In terms of the internal market demand (IDM), the environment Kuznets curve suggests that environmental pollution and per capita revenue are in the form of a "U" relationship, with only a critical or "turning point" per person per capita, environmental pollution will be reduced by the increase in per capita incomes, and the environmental quality of the environment is improving. The income level of urban residents is negatively correlated with the green innovation performance, indicating that our country is still in the left of the inflection point of the environmental Kuznets curve, and the current development model of pure pursuit of quantity growth has led to the lack of attention and even neglect of green innovation in the process of economic development in China. In terms of external market demand (EDM), some export enterprises excessively pursue the expansion of export scale due to lack of sufficient environmental protection awareness, ignoring the environmental impact and resource consumption caused by the export. Driven by the maximization of economic benefits, export enterprises tend to obtain more economic profits at the expense of environmental benefits, and reduce operating costs by reducing or even avoiding the investment of green innovation, which is not conducive to the improvement of green innovation performance. Thirdly, in terms of institutional factors, the regression coefficient of intellectual property protection system (IPS) and market system (MS) is significantly negative, which also has negative influence. In the case of IP protection, the stricter intellectual property protection can effectively prevent "hitchhike" behaviour, increase the initiative of innovative subject green initiatives; At the same time, however, the strict intellectual property protection reduces the possibility of the innovative body to acquire the external advanced green information, and increases the simulation of innovation difficulties and prevents the diffusion of green innovation results. So far, the green innovation in our country is based on imitation of innovation, the ability to innovate autonomously, and strict intellectual property protection has restricted the diffusion of green innovation, and it's bad for the development of green innovation in the region. In view of that market system (MS), the specific gravity of state-owned enterprise reflects the degree of monopoly in the market, and the higher the proportion of the state, the more the monopoly is, the lower the efficiency of the green innovative resource allocation, and the lower the effect of green innovation. In addition, state-owned enterprises are affected by the diversification of targets, which leads to insufficient attention to green innovation activities and insufficient incentive for green innovation. Therefore, the higher the proportion of state-owned enterprises, the less conducive to green innovation activities and the improvement of green innovation performance.

The regression coefficient of the environmental system (ER) is not significant, while the environmental system ER(1) is significantly positive, which means that the environmental system has a positive influence on the one year lagged regional green innovation performance. Compared with the traditional innovation, green innovation can achieve multi-party benefits of economy, society and environment, but the environmental benefits brought by green innovation will be acquired by the society, while the green innovation subject should pay more cost than traditional innovation, which leads to the lack of sufficient innovation incentive for green innovation. Therefore, if no external constraint is applied, the competition between green technology and traditional technology will be distorted, thus reducing the enthusiasm of regional green innovation, which is unfavourable to the improvement of green innovation performance. It is shown that that environmental control strength has significant positive influence on green innovation performance in our country, and the better the green innovation performance in the region with higher environmental control intensity, which supports the "Potter Hypothesis" theory. In order to further compare the regional difference of the above factors influencing on the green innovation performance, a regression estimate is carried out from three sub-samples in the east, central and west.

Firstly, technical factors. Foreign direct investment (FDI) in the eastern region of the green innovation performance has a significant negative influence. This difference mainly lies in the gradual decline of the green innovation ability in the three regions. The green innovation capacity and production technology of most provinces in the eastern region are quite than that of foreign enterprises. It is difficult to obtain the spillover effect of green innovation by means of model imitation effect and association effect. However, the competitive effect caused by foreign enterprises will aggravate the vicious competition among enterprises in the eastern region, which will not be conducive to the improvement of green innovation performance in the region. In the western region, the green innovation ability and production technology environmental protection level are relatively low. In recent years, the transfer of foreign capital to the western region has enabled the region to obtain spillover effects through demonstration imitation and industrial association, and to promote the improvement of green innovation performance. External outward direct investment (ODI) has a significant positive influence on green innovation performance in the central and western regions, but has no significant impact on the eastern region. The central and western regions can improve the performance of green innovation through the reverse technology spillover of FDI. Secondly, the internal market demand (IDM) has a significant negative impact on the green innovation performance in the central and western regions, and has a more significant impact on the western region, while the impact on the eastern region is not significant. It's a feature that fits the environmental Kuznet curve, which means that income of urban residents of the eastern part pays much more than the central region, and the central region pays more than the west. The external market demand (EDM) has negative influence on the green innovation performance in the eastern region, which has no significant influence on the central region, and has a significant positive influence on the western region. This is mainly because of the central region export-oriented industrial enterprises were significantly higher than those in the western region, and the eastern region export enterprises will lead to the fierce competition between enterprises which are more pursuit of scale expansion with low cost, and decrease the cost of green innovation into further compression. Thirdly, institutional factors. The intellectual property protection system (IPS) has a positive impact on the green innovation performance in the central and eastern regions, but the regression coefficient has not passed the significance test, while in the western region is significantly negative. This is also because of the high green innovation capability in the eastern and eastern regions, while the green innovation capability in the western region is lower. And, the more strict intellectual property protection system reduces the possibility of model innovation in the western region, which is not conducive to the improvement of green innovation performance in the region. The market system (MS) has a significant negative influence on the green innovation in the eastern region, while the negative influence on the central and western parts is not significant. Besides, the green innovation ability in eastern regions is much higher than the other regions, but still needs to build more market-based competitive mechanisms to increase the efficiency of green innovation, so the green innovation is restricted by the state-owned enterprises. The regression coefficient of the environmental system (ER) was not significant in all three regions. However, the regression coefficient of ER(1) is significantly positive in the central and eastern regions, while the western region is not significant. So the one year lagged environmental regulation has a positive influence on the green innovation performance in the central and eastern regions.

4. Conclusion

In terms of green innovation performance evaluation, the mean value of regional green innovation performance evaluation showed the M type volatility tendency in the whole country in 2005-2014. The green innovation performance evaluation in the eastern region has the highest value, and the central is the same as the national average, and the evaluation value in the western is significantly lower. Therefore, in order to reduce the green innovation performance gap in the eastern and western regions, we need formulating the moderate policy in different regions. Local governments in the central and western regions must increase the local green innovation fund support. Through tax incentives and financial subsidies lead the central and western enterprises to carry out green innovation. To improve the financing system and environment of green innovation, we need to channel more social capital into green innovation investments. At the same time, we should establish a strict system of intellectual property rights protection, protect the excess profits brought by the innovation of green innovation, and avoid the incentive of free riders to reduce the innovation enthusiasm of green innovation subjects.

In terms of influencing factors of the whole nation, foreign direct investment(FDI), outward direct investment(ODI), the one year lagged environmental system has a positive impact on the green innovation performance, while internal and external market demand, the protection of intellectual property rights system, market system has a negative influence on it. From the perspective of the three regions, FDI and external market demand have a negative impact on the green innovation performance in the eastern region, and have a positive influence on the western region. The foreign direct investment has a positive impact on the green innovation performance in the central and western regions. The internal market demand has a negative impact on the green innovation performance in the central and western regions. The intellectual property rights protection system has a negative impact on the green innovation performance in the western region. The market system has a negative impact on the green innovation performance in the eastern region, and the one year lagged environmental system has a positive influence on the green innovation performance in the central and eastern regions. Therefore, when considering foreign direct investment, outward direct investment, external market demand and other open economic factors, we should focus on the development of east, middle and western regions. In introducing foreign direct investment, we should pay attention to the level of foreign investment in environmental technology, but also pay attention to the leading principle of "sub-region and classification", that is, according to the economic development priorities and development level of the three regions of eastern and western China, we will formulate a catalogue of green technologies that are suitable for the local economic development and absorption capacity, and do a good job in the distribution of echelons in the three regions. In formulating environmental control policies, we should establish a decentralized environmental control model that is more suitable for economic development, and give local standards under the premise of national uniform standards. Improving the market competition mechanism can direct the green shift of market demand. On the one hand, we will continue to deepen the reform of state-owned enterprises, invigorate the market, and improve the allocation of green innovative resources through market mechanisms; On the other hand, the promotion of green consumption concept and education can strengthen the green consumption awareness among enterprises and consumers, and lead and support the social purchase of green products to form a good green consumption pattern.

5. References

[1] Cheng H, Liao Z J.,2011. China regional environmental innovation performance evaluation and research. China Environmental Science. 31(3):522-528

[2] Bi K X, Yang C J, Huang P.,2013. Study on the regional difference and influencing factors of green process innovation performance in China. China Industrial Economics. 30(10):57-69.

[3] Wang H L, Lian X Y, Lin D M.,2016. Effects of Green Technological Innovation Efficiency on Regional Green Growth Performance: An Empirical Analysis. Science of Science and Management of S.& T. 37(6): 80-87.

[4] Cheng C C, Shiu E C.,2012. Validation of a proposed instrument for measuring eco-innovation: An implementation perspective. Technovation. 32(6): 329-344.

[5] Huang X-X, Hu Z-P, Liu C-S, et al.,2016. The relationships between regulatory and customer pressure, green organizational responses, and green innovation performance. Journal of Cleaner Production. 112(4): 3423-3433.

[6] Fan Q L, Shao Y F, Tang X W.,2013. The impact of environmental policy, technological progress, and market structure on environmental technology innovation. Science Research Management. 34(6):68-76.

[7] Shen N, Liu F C.,2012. Can Intensive Environmental Regulation Promote Technological Innovation ? : Porter Hypothesis Reexamined. China Soft Science. 27(4): 49-59.

[8] Zailani S, Govindan K, Iranmanesh M, et al.,2015. Green innovation adoption in automotive supply chain: the Malaysian case. Journal of Cleaner Production. 108(3):1115-1122.

[9] Fan Q L, Shao Y F, Tang X W.,2013. The impact of environmental policy, technological progress, and market structure on environmental technology innovation. Science Research Management. 34(6):68-76.

[10] Cuerva M C, Triguero-Cano Á, Rcoles D.,2013. Drivers of green and non-green innovation: empirical evidence inLow-Tech SMEs. Journal of Cleaner Production.68(2): 104-113.

[11] Yang C J.,2014. Impetus evaluation and regional difference analysis of green process based on RAGA-PPE model. Science & Technology Progress and Policy. 31(18): 51-56.

[12] Li S L, Shen C, Lin P N.,2014. Environmental regulations and regional economic growth ——based on the empirical test of China's provincial panel data. Collected Essays on Finance and Economics.30(6): 88-96.

[13] Yang R F.,2015. Industrial agglomeration, foreign direct investment and environmental pollution. Business Management Journal. 37(2): 11-19.

[14] Li W H.,2015. Spatial econometrics test of pollutant discharge system’s driving on green technological innovation by taking 29 provinces and regions’ manufacturing industries as examples. Science Research Management. 36(6): 1-9.

Supported by the National Natural Science Foundation of China (Grant No. 71502074)

IOP Conf. Series: Earth and Environmental Science **182** (2018) 012019 doi:10.1088/1755-1315/182/1/012019

Development of Bicycle Transport in the City of Sofia as Part of the Concept for Stable Urban Mobility

S D Tzvetkova

University of National and World Economy, Sofia, Bulgaria
E-mail: svetlatzvetkova@abv.bg

Abstract. The concept for Stable Urban Mobility, included in the European Commission's Green Papter titled "Towards a New Urban Mobility Culture", consists of encouraging the combination of different types of public transport with different types of individual transport. The new concept for urban mobility also suggests reaching common goals for economic prosperity and recognizing the right to mobility by managing the quality of life and protecting the environment. The Green Paper states the following: "Encouraging walking, cycling and constructing the appropriate infrastructure" which makes it clear that bicycle transport and its related infrastructure are an inseparable part of stable urban mobility. However, the city of Sofia presently does not have a developed bicycle infrastructure. Incorporating bicycle traffic within the city's transport framework is not only necessary, it has to become a part of the priorities of the city's transport system. The purpose of the report is to substantiate the necessity for the development of bicycle transport in Sofia and to indicate the guidelines for its future development. The more stably built an urban transport system is and the more corresponding it is to citizens' needs and to the problems of modern urbanized environments, the higher the guarantees are for the city's stable development in society's favor.

1. Introduction

The large number of automobiles, the dynamic of urban traffic and the increased traffic worsen the quality of the urban environment and are the main cause for air pollution in the city of Sofia. About two thirds of the city's population travels with their own personal cars or by buses, which are slow, depreciated and harmful to the environment. The city also has trolley and tram transport but due to the poor maintenance of the tram infrastructure and the decreasing number of trolleys, these types of transport handle only 25% of the passenger flow of the capital's ground urban passenger transport. Sofia's polluted air is one of the main reasons for the growing health issues – namely, respiratory and cardiac problems which start from an early age. Bulgaria holds second place when it comes to the number of fatalities caused by cardio-vascular diseases. In recent years Sofia has become one of the most air-polluted cities with harmful emissions and noise pollution. All of this also has a negative impact on social and economic development. It is apparent that environmental policies have never been a priority in the capital's agenda despite Bulgaria's membership in the EU, which vehemently demands that its member-states adopt stable development practices. It is only recently that Sofia has started a policy for reducing the harmful influence of urban transport on the environment; designing a program for the development of bicycle transport is part of said policy. In order to reduce traffic jams and improve the city's traffic quality, bicycle traffic has to become part of the urban transport system [1]. Therefore, it is imperative that the city of Sofia provide the necessary conditions for it and construct bicycle lanes along with a stable urban transport infrastructure. It is crucial that the city become a modern "green" European city with a healthy and high-quality urban environment which will allow people of all ages to move safely, quickly and comfortably on bicycles. The appealing urban environment will help people become more positive, happy and able-bodied. In turn, the high-quality urban environment will atrract more investments and tourists to the city. In practice, increasing bicycle traffic will improve conditions for the other transport traffic participants [2]

2. Analyzing Bicycle Traffic in the City of Sofia

2.1. Data for Using a Bicycle in the City

The main indicator for the dynamic and the mobility in every city is the sorting of travels by different types of vehicles. Figure 1 presents data from a survey done by experts when designing the "Program for Developing Urban Transport Within the Sofia Municipality for the 2016-2019 Period" [3]. The data is about the necessity for determining the percentage of people who use public transport, personal vehicles, bicycle transport and people who prefer to walk. The survey has been carried out within the city of Sofia for 2011 and 2016.

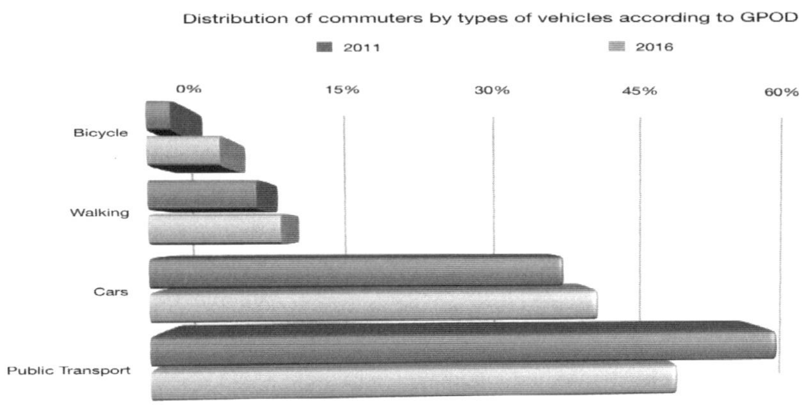

Figure 1. Sorting travelling people by types of vehicles

Conclusion: The figure's presentations lead to the conclusion that the percentage of people who use bicycle transport in 2011 is only 2%. It could also be acknowledged that the rate of increase for 5 years is too small – in 2016 it is only 6%. As the figure clearly indicates, the city of Sofia displays an extremely negative trend for the development of bicycle transport.

2.2. Analyzing Bicycle Ownership

In 2016 the "Veloevolution" Association carried out a survey among 2536 people to determine the number of bicycle owners in the city of Sofia. The survey data shows that bicycle owners comprise only 18.1% of 1000 households or 7% of the interviewees, with most households owning only one bicycle each. Of all the vehicles that citizens use, bicycles comprise only 24.7%, but the majority of them are used only during weekends and holidays.

Conclusion: The percentage of citizens who own a bicycle is extremely low due to the lack of the necessary conditions and they see no point in owning one. The majority of citizens are concerned about the poor road conditions and the probabilities for traffic accidents caused by other traffic participants. Statistics show that in 2016 the city of Sofia has registered a total of 1064 traffic accidents, with 42 fatalities and 1311 injured citizens. The number of accidents involving cyclists for the same year is 444, with 35 fatalities and 409 injured. The majority of the casualties are aged 25-65 [4]. This is an extremely negative statistic, considering the small percentage of people who use bicycles to travel in the city.

2.3. Reasons for Which Citizens Do Not Use a Bicycle

The main reasons for which people do not cycle are the dangers that could arise on the road, as well as thefear of bicycle theft [3]. The increased number of traffic accidents in the city of Sofia, the poor condition of the road infrastructure and the inappropriate behavior of the other traffic participants endanger the lives of cyclists (see fig.2).

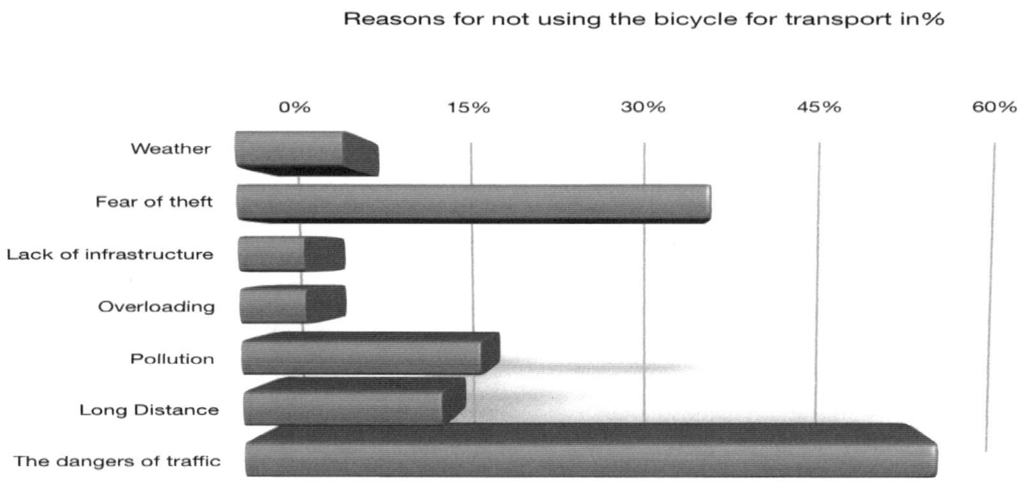

Figure 2. Reasons for which citizens do not use bicycle transport

Data: "Veloevolution" Association, based on a survey carried out among 143 people in 2016

Note: The analysis is based solely on active cyclists and does not examine the problems of non-cyclists

Conclusion: In order to satisfy cyclists' needs, the first priority should be the implementation of adequate measures for differentiating bicycle road-beds and constructing secure bicycle parking lots to improve citizens' culture and tolerance for cyclists.

2.4. Reasons for Which Citizens Would Start Traveling by Bicycle

Over 60% of the residents declare that they would not cycle even if travel conditions are improved (see Table 1). It is interesting to note that in Bulgaria cycling is viewed more as a hobby or a sport, whereas Western European countries such as Germany and the Netherlands have a built-in bicycle culture, cycling there is something normal and people use their bicycles every day, not just on weekends and holidays, and they cycle to work.

Table 1. Reasons for which citizens would start traveling by bicycle

Type of reasons	*% of citizens*
If separate bicycle lanes are constructed	17.9 %
If bicycle lanes are differentiated as part of the roadway	15.1 %
If cycling conditions in the city are improved	11.9 %
If the necessary conditions for cycling to work are provided	7.9 %
If safe cycling courses are organized	3.0 %
I would not cycle in the city under these conditions	63.7 %

Data: "Veloevolution" Association, 2016, a survey carried out among 2536 people

Conclusions: Unfortunately, cycling is extremely unpopular in Bulgaria. Therefore, comprehensive measures need to be taken to increase road safety and especially boost the image of cycling and popularize it through investing in the construction of a bicycle infrastructure, organizing various cycling events, media relationships for more coverage, articles about cycling, politicians and the municipality setting a personal example among employees about cycling, etc [3].

2.5. Preferences of Citizens for Priority Bicycle Infrastructure

The figure's data (see. fig. 3) shows the desires and expectations of cyclists. The main issue for cyclists is the lack of bicycle lanes. The indication for the need of bicycle lanes from the neighborhoods to the center clearly shows people's willingness to cycle to work or to school. People who ride their bicycles in parks or for sport also point out the lack of bicycle lanes and their desire for those [5].

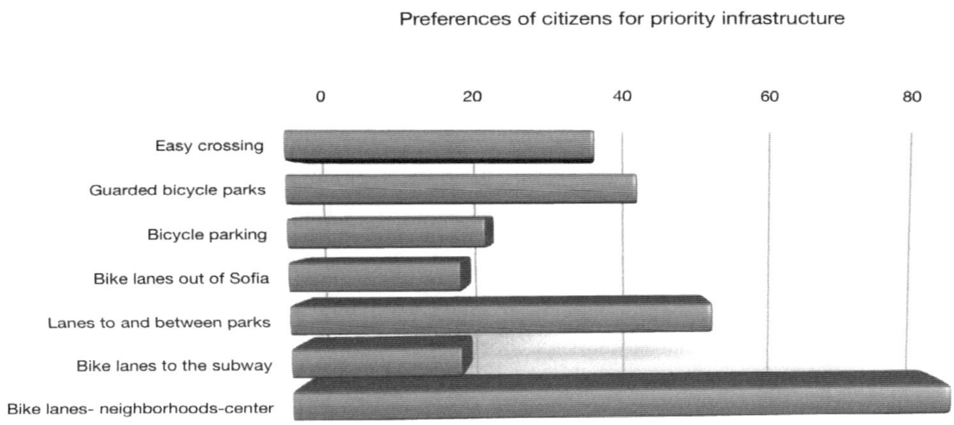

Figure 3. Preferences of citizens for priority infrastructure
Data: "Veloevolution" Association, a survey carried out among 143 people, 2016

Conclusion: The construction of a primary bicycle network from the neighborhoods to the center as well as bicycle crossings at busy intersections should be a priority. Connecting the green areas in the city with bicycle road-beds will help them function as a system for improving the possibilities for bicycle rides in the city.

3. Recommendations for the Stable Development of Bicycle Transport in the City of Sofia

3.1. Building a Bicycle Infrastructure

The construction of new bicycle lanes and the maintenance of the ones that already exist, as well as the environment around them, should be priority number one for the future development of bicycle transport in Sofia. In order to guarantee safety, reduce travel time and improve cyclists' comfort of movement, it is necessary to improve the infrastructure [6]. The city of Sofia does not have bicycle road-beds from the big residential districts to the center, or a bicycle ring around the center of Sofia – for faster movement. An indicative example in this regard is the city of Münster, Germany. The availability of a park with bicycle lanes in the inner city allows its citizens to move about undisturbed by automobile traffic. This will reduce the risk of accidents and increase the number of cyclists. In addition, it is neccesry to connect bicycle routes to subway stations, i.e. provide an opportunity to combine the bicycle transport and the subway as well as construct guarded bicycle parking lots in the city and at subway stations. Since bicycle transport is an inseparable part of bicycle infrastructure, the availability or the lack of bicycle parking spaces can have a significant impact on people's choice to cycle. Designing a separate plan for bicycle parking is a good practice, especially for larger cities, in order to help improve the developent of the spaces and the capacity for bicycle parking lots and maintain an actual map of parking spaces. This will provide an opportunity to track the usability of bicycle parking lots and correct the number of parking spaces when necessary. Information about bicycle parking lots could be inserted in printed or interactive bicycle maps. A good bicycle lane network in the center is also a necessary start for the establishment of a system for public bicycles.

3.2. Developing a System for Public Bicycles

Public bicycles are individual means of public transport. They are bicycles which are usually placed at special stations throughout the city and are only available to every citizen after a specific registration and/or payment, depending on the chosen system [7]. Denmark, Germany and the Netherlands are typical examples. They complement the mass public transport network by allowing people to reach places that are inaccessible for large vehicles. The positive influence of introducing public bicycles has helped people who do not own a bicycle or live far from the center start cycling. Mass urban transport becomes increasingly appealing when public bicycles help traverse the relatively short distances which, however, take far longer with motorized transport due to traffic jams. A campaign for the establishment of a system for public bicycles has to be large-scale enough in order to have a bigger impact on citizens and attract more customers. Based on a study of the systems in Paris, Barcelona, London, Vienna, etc., it is advisable to start with no less than 100-120 bicycles placed at about 10 stations. When it comes to the desire for establishing a system for public bicycles, a thorough survey should be carried out, and the public opinion and the task should be analyzed. Before establishing the system, informational campaigns should be organized to make citizens aware of the service. The introduction of public bicycles should be done after there has been a minimum conscription of bicycle lanes in the city. Usually the system for public bicycles starts at the center of the city where the usability of bicycles is the largest. Subsequently, the system could extend to other parts of the city as well.

3.3. Building Suburban Bicycle Routes

Sofia's proximity to five mountains, two dams and the cultural landmarks in the city's suburbs provides big opportunities for developing suburban tourism. Creating commodities for cycling to these destinations is an alternative to driving – a new experience and active vacation which citizens need, considering the sedentary lifestyle. Even people who have little cycling experience would be able to try cycling – during a holiday, on a comfortable and safe road-bed. This could enhance their cycling skills and encourage them to use bicycles as a means of transportation in their everyday travels [3]. In order to stimulate bicycle tourism and stable transport, facilitations for cycling in the suburban routes of public transport should be introduced. Providing more commodities and opportunities for cycling out of town will stimulate more people to prefer it for their vacation. Whether it is organized or individual, the opportunity for access to cultural landmarks by bicycle will be a huge advantage to the city as a tourist destination. For this purpose, it is necessary to build bicycle routes between cultural landmarks and improve the road-bed infrastructure, especially at intersections and crossings. Bicycle stands should be installed near the entrances to the sites, routes should be marked, informational panels with maps should be available and printouts of the city's bicycle map should be distributed. The practice of renting bicycles should be encouraged, seeing how it is commonly accepted in Western countries.

3.4. Integrating Bicycle Transport into Public Urban Transport

Transporting bicycles in ground urban transport is not explicitly regulated in the capital's mass urban transport. Therefore, it is necessary to adopt certain measures in that regard. Bicycles could be relegated to chargable oversized hand luggage and it has to correspond to the condition that it will not pollute or damage the vehicle, nor will it create inconveniences for the other passengers. The transportation of bicycles in ground urban transport should be permitted under conditions similar to the ones in the subway and with limitations on the number of bicycles allowed aboard one vehicle. Railway transport and bicycle transport can also be combined by changing the vehicle at railway stations. Bicycle parking lots should be installed at passenger railway stations and stops in the city of Sofia, as well as at railways in other populated areas which are exit points for everyday trips to Sofia. It is advisable that bicycle parking lots be guarded or at least situated within the station or near its entrance, like in Amsterdam and Münster. As a main "entrance" to the city, the central railway station should have a covered guarded bicycle parking and it should be suitable for the creation of bicycle road-beds from the stations to the city, which will facilitate and stimulate people to combine bicycles with railway transport. In Germany transporting a bicycle aboard a train is something ordinary; there is even a special car solely reserved for bicycles. Creating facilitations for combining mass urban transport with bicycles makes the use of public transport more appealing, more flexible and more competitive to driving a car and, at the same time, it encourages people to ride bicycles.

3.5. Measures for Improving Road Safety for Bicycle Transport

Ciclysts are very dependent on the other traffic participants and their culture. About 75% of traffic accidents involving cyclists end with injuries and fatalities, whereas with other accidents they are about 10%. Additionally, road danger is the main reason why people who own bicycles do not use them as an everyday means of transportation. Statistics in other countries indicate that the higher number of cyclists on the road reduces the number of cyclist fatalities per km traveled distance. The reason is that drivers become more cautious in the presence of multiple cyclists. Road safety for cyclists can be improved by: creating zones with a speed limit of 30 km/h – in residential districts and in the center; improving dangerous intersections and crossings – for example, creating a stop line for cyclists placed ahead of the stop line for motorized traffic; adding bicycle lanes to cross-walks; extending the road surface for bicycle lanes across intersections if the bicycle lanes are major; creating intermediate refuges and expanded sidewalks when crossing; narrowing the roadway and reducing the radius of curb stone curves; controlling the observation of traffic rules by all traffic participants; sanctioning the other traffic participants if they endanger the lives of cyclists; improving road safety education and training in schools, driving courses and exams[8]. Very often cyclists feel most endangered at intersections and crossings. Some places in the city of Sofia are stull difficult to cross and with very high risks for cyclists. Introducing basic traffic and road safety rules to children is essential to providing them with permanent discipline and enhanced transport culture. In Bulgaria road safety training is mandatory in schools. Multiple companies and NGOs also organize road safety campaigns, especially for children. In addition to training children at school, special road safety training should also be given to future drivers. In the process of training future drivers, special attention should also be given to actions wih regard to the presence of cyclists and the bicycle infrastructure on the road – i.e. predicting their behavior and respecting their privileges. Aside from the ability to ride a bicycle, cycling in urban conditions poses challenges for which cyclists should also be prepared. It is highly advisable that cyclists be as familiar with road traffic rules as drivers of motorized vehicles. Observing the basic rules for road safety and behavior on the part of cyclists would reduce the number of road conflicts and accidents.

3.6. Awareness and Communication with Citizens

The city's image, the environment's quality, the manner of using public spaces, etc. result from complex interactions and making political and expert decisions, from settlements by compromise with all interested parties, from financial possibilities, communication with citizens, etc. [9]. However, in every step several options should be taken into consideration in order to reach better decisions. In recent years transparency and communication with citizens have become especially important. The successful bicycle policy is expressed in the public internet space as a separate page or section in the webpages of the municipality or the public transport authorities, where plans for a new bicycle infrastructure, reports and analyses of bicycle traffic, sugestions and signals from citizens regarding bicycle infrastructure, etc. can be published. Creating offices for informing citizens about the capabilities of urban mobility is a good practice. The high quality of designing and executing bicycle infrastructure and urban furnishing influences the quality of the entire urban environment. Therefore, it is important to create a unified "cycling" style when designing bicycle infrastructure, road surfaces, details, signalzation, signature, signs for orientation, bicycle stands and parking lots, etc.

3.7. Actions for Making a Successful Bicycle Policy

The measures for encouraging bicycle traffic go hand in hand with: measures for reducing automobile traffic; determining a budget for bicycle infrastructure, management and monitoring; events like a

European Week of Mobility and other undertakings which can prove successful in coordination with various actions for establishing a new urban mobility culture. It is appropriate to form a bicycle commission to discuss current cycling issues in the city – such commissions function in several European citires like Wrocław, Vienna, Bratislava, etc. A qualified team dealing with bicycle infrastructure has to make citizens aware of innovations as well as the good and bad practices in the area. The policy for urban development planning has to be inextricably bound to the desires and security of cyclists. When designing every new project for new construction, reconstruction or a major overhaul of streets or public places, the passing of cyclists, even if it is not included in the bicycle plan, should also be taken into account. The simultaneous construction of streets and bicycle lanes will spare authorities money and inconveniences if the addition of a new bicycle lane becomes necessary subsequently.

4. Conclusion

The ever-growing number of automobiles in the city of Sofia often makes city traffic difficult and significantly increases travel time. Therefore, many people would choose bicycles as an alternative means of transportation. Bicycles could become a reliable transport, especially for traveling across the busy central part of the city. The city of Sofia holds significant potential for developing urban and suburban bicycle routes. Very often people own bicycles but they use them only during vacations due to the lack of a built-in infrastructure and conditions for safe movement. If new bicycle lanes are built and the ones that already exist are maintained well, this will attract more citizens to this way of traveling and it will create prerequisites for the stable development of bicycle transport in Sofia. An urban bicycle network should be integrated with the other types of public transport in order to achieve comfort and effectiveness when using bicycles as a conductive means of transport to subway stations and stops. If the accessibility by bicycle to cultural landmarks in the city and around it is improved, the city will become a more appealing tourist destination, which is especially significant within the context of Sofia being a candidate for cultural capital of Europe in 2019[10]. Cycling improves the health condition of the city's residents, especially children, and it helps preserve the environment. The development of bicycle transport is in the context of creating stable urban mobility for the EU's member states and it should become a priority in the policy for the future development of the city of Sofia.

5. References

[1] Strategy for developing the transport system of the Republic of Bulgaria by 2020, p. 40;
[2] Plan for the development of Sofia Municipality 2007-2013;
[3] "Program for the Development of Bicycle Transport Within the Sofia Municipality for the 2016-2019 Period", pp. 23-27;
[4] National Statistics Institute – 2017, "Traffic Accidents in the Republic of Bulgaria for 2016";
[5] "Veloevolution" Association;
[6] "Program for the Development of Bicycle Transport Within the Sofia Municipality for the 2012 -2017", pp 45-47;
[7] Sofia Municipality's "General Plan for Traffic Organization";
[8] Strategy for improving road and traffic safety in the Sofia region, 2012-2020, pp. 17-19;
[9] Modification of the Common Spatial Development Plan of SM – 2009;
[10] Velikova, E (2017), "Trends in the planning and construction of the tourist destination", Jornal Real Estate Property & Business, Vol.1(1), pp. 41-45.

Comparison of the Cyclic Variation of a Diesel-Ethanol Blend in a Diesel Engine

M H Mat Yasin[1], A F Yusop[2], R Mamat[2], A A Abdullah[2] and N H Badrulhisham[2]

1 Department of Mechanical Engineering, Politeknik Kota Kinabalu, 88460 Sabah, Malaysia
2 Faculty of Mechanical Engineering, Universiti Malaysia Pahang, 26600 Pahang, Malaysia
Email: hafizil@polikk.edu.my

Abstract. Alcohols are renewable and sustainable second generation biofuels which are derived from various biomass feedstock sources. These fuels with similar properties to mineral diesel can be used as a blend or additive to improve the combustion characteristics and pollutant emissions in the automotive engines. However, different fuel properties characterize different combustion phasing parameters for the specific engine operation and test condition. This paper presents the preliminary results of coefficient of variations of IMEP (COVIMEP) and P_{max} ($COVP_{max}$) for a diesel engine fuelled with mineral diesel (B0) and DE10 blend at full load both engine speeds of 1100 rpm and 2300 rpm. The influence of ethanol content in a blend of diesel on the cyclic combustion variations is explained in the calculation values of the coefficient of cyclic variation (COV). The experimental results showed the DE10 fuelling exhibited larger cyclic variations than mineral diesel (B0) at the same test conditions, owing to the reduction of combustion temperature during combustion phasing and lower reactivity of ethanol.

1. Introduction

Alternative fuels have potential to go further to improve the compression ignition engines energetically and pollutant emissions. Alcohols including ethanol and butanol represent second generation biofuels which score high ratings in combustion properties, sustainable feedstock sources and possibly diminish the fossilized fuel consumption [1–3]. Therefore, ethanol is considered an alternative fuel for the automotive engines at present and could reduce the harmful emissions. A 40% - 60% reduction in the NO_x emission is found with the use of ethanol. Ethanol is considered an alternative sustainable fuel for automotive engines due to its several advantages including comparable properties with SI engines at specific operating conditions [4,5], produced from agricultural and waste products through chemical processes [6,7] and infrastructure for present fuel able to facilitate the distribution and storage possibilities [8]. Ethanol possesses better combustion properties comparative with the present fuels with higher octane number [9], more substantial oxygen content at the molecular level [8], greater autoignition temperature [10], greater laminar flame speed [11] and lower adiabatic flame temperature. There is no significant engine modification of its design needed when running with the ethanol in maintaining or increasing the engine performance energetically. Since ethanol has a higher heat of vaporization, an improved intake air efficient cooling effect is obtained which leads to a volumetric efficiency improvement and reduces the risk knock development. Also, pollutant emissions especially NOx is decreased owing to the lower in-cylinder temperature. Moreover, the use of diesel-ethanol blends leads to the increase of the maximum in-cylinder pressure and the pressure rise maximum rate due to the advanced combustion properties.

Cyclic variations determine the characterization of the engine when running with different fuel and test operating condition. Numbers of research work have focused the investigation on the cyclic variations, then proposed several techniques for measuring and analyzing the cyclic combustion variations in spark ignition (SI) engines [12]. Most of the studies are focused on reducing the SI engines unpredictability, particularly in engine knocking. However, combustion cyclic variation investigations recently have been conducted on standard diesel engines by other researchers operating with different fuels including alcohols [13]. The cyclic variations can be characterized by coefficients of variation (COV) in cylinder pressure. The intensity of the cyclic variation occurrence is defined by the coefficient of cycle variations for some specific cycles. The coefficient of cyclic change is described as a relative average deviation of maximum pressure values in the engine operation. For "n" consecutive cycles, if is considered a normal distribution of the deviation probabilities, the squared average standard deviation, σ can be calculated and the cycle variation coefficient (COV) is defined as:

$$COVa = \frac{SD(a)}{\bar{a}} \times 100\% \quad (1)$$

Where,

a is the parameter of which variability is studied and is defined as for indicated mean effective pressure (IMEP), maximum pressure (P_{max}) in the cycle number "i".

SD (a) is the standard deviation of the defined parameter

IMEP is the mean value of the defined parameter

$$SD(a) = \sqrt{\frac{\sum\left(a_i - \bar{a}\right)^2}{n_c}} \quad (2)$$

$$\bar{a} = \frac{1}{n_c} \sum_{i=1}^{N_c} a_i \quad (3)$$

Where,

a_i is the pressure at specific cycle (bar)

n_c is the number of cycles

In general, the way of cyclic variation evaluation for test condition closer to the value of maximum brake torque the COV of maximum pressure, P_{max} is suitable. The variation of the IMEP, translated by the coefficient of variation (COV) IMEP, is the most suitable instrument to characterize the engine respond to the differences in the combustion process. It is precisely shown that the limit value of COVIMEP defines strictly the limit of mixture leaning. Also, this coefficient of cyclic variation can also point toward the difference in flame development during the initial combustion phase. Hence, the fuel type used in the engine influences the cyclic variation as a result of its laminar flame velocity values. Therefore, the flame development of ethanol is much quicker, comparative to gasoline and diesel owing to the higher laminar combustion speed. This attribute reduces the influence of turbulence and the cyclic variations. However, the quality of the in-cylinder mixture influences the combustion process through chemical reaction speed, with a maximum is obtained in the area of rich region. From this point of view, the initial and final phases of the combustion process have a minimal duration in the region for which the chemical reaction speeds are maximum. At the mixture leaning the durations of those two phases increase and the total combustion duration also increases.

2. Experimental Set up

In this research work, mineral diesel (B0) was purchased from a local supplier company, and ethanol was purchased from Merck through a local agent. Then, the fuel was blended at 10% by volume of ethanol for each 1 liter of mineral diesel (B0), which denoted as DE10 (90% mineral diesel+10% ethanol). The fuel properties of the blend are tested according to the ASTM standards for density, viscosity, cetane number, flash point and calorific value. Table 1 lists the test fuel properties in the study.

Table 1. Test fuel properties

Description	Testing Method (ASTM)	Mineral diesel (B0)	DE10
Density @ 20 °C g/cm³	D287	0.8264	0.8226
Viscosity @40 °C mm²/s	D445	5.144	3.674
Cetane number	D613	47.8	43.82
Flash Point (°C)	D93	60	55.2
Calorific value (MJ/kg)	D240	44.8	43

This research work was conducted on a Yanmar TF120-M four stroke, direct injection single cylinder diesel engine. It is a water-cooled, low-speed with a maximum power of 7.8 kW at 2400 rpm. A 15 kW eddy current, dump load dynamometer was coupled to the engine and been controlled with a universal controller model DC5-10KW for controlling the engine speed and torque. Two separate fuel tanks with thermocouples and a fuel valve system were used, one for mineral diesel, B0 and the other for the DE10 blend. In fuel delivery system, a burette was used to measure the fuel consumption of both fuels. Table 2 describes the specification of the test engine and Figure 1 illustrates the setup of the engine testing.

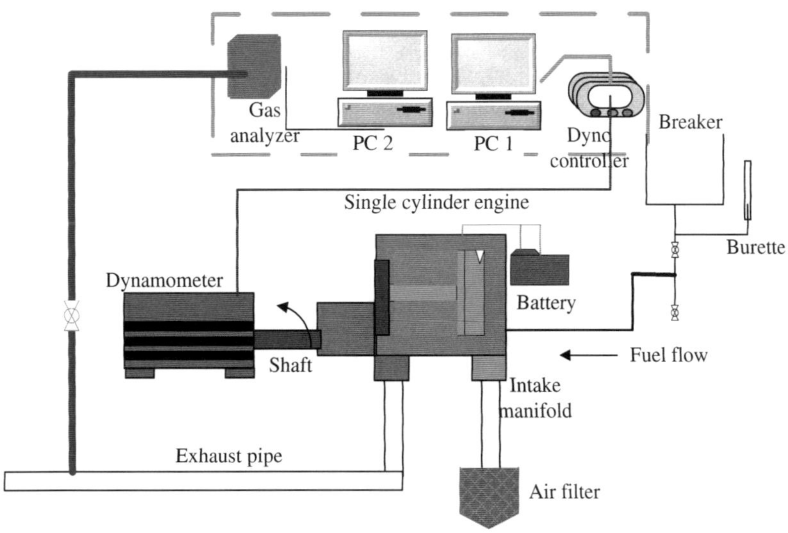

Figure 1. Engine testing set up

A preliminary comparative fuel study of cylinder pressure cyclic variations was developed for full load condition and both engine speeds of 1100 rpm and 2300 rpm. Therefore, the in-cylinder pressure was measured and recorded using an Optrand AutoPSI-S model C22294-Q pressure transducer with a measurement range from 0-5000 psi. The crank angle degree signal was obtained using a magnetic crank encoder. A TFX combustion analyzer was used to record and analyze the in-cylinder pressure and crank angle signal measurement at the specific test condition as listed in Table 3.

Table 2. Test engine specification

Description	Specification
Engine model	Yanmar TF120M
Number of cylinders	1
Combustion system	Direct injection
Total displacement (L)	0.638
Bore x stroke (mm)	92 x 96
Injection timing	18° BTDC
Compression ratio	17.7:1
Continuous output (HP)	10.5 HP at 2400 rpm
Rated output (HP)	12 HP at 2400 rpm

Table 3. Test condition

Parameters	Test condition
Type of fuel	B0, DE10
Speed (rpm)	1100, 2300
Load (%)	100
Fuel temperature (°C)	27 ± 1
Air temperature (°C)	30 ± 1

These in-cylinder pressure data were selected to ensure the data consistency and were collected for 200 consecutive cycles for both B0 and DE10 fuelling. In this research work, a further calculation is needed to obtain the cyclic variation coefficients for indicated mean effective pressure (IMEP) and maximum pressure (P_{max}). To evaluate the way that the engine running is affected by the variability of the combustion process, these coefficients are calculated and presented in the following figures. Regarding cycle variability, the general tendency shows a significant decrease in this phenomenon when the DE10 fuel is used.

3. Results and Discussion

For all running test condition, full load and both engine speeds, the injection timing was maintained to ensure its reliability. At an operating test condition, defined by engine speed and full load, the fuel cycle dose was continued using the ball valve opening.

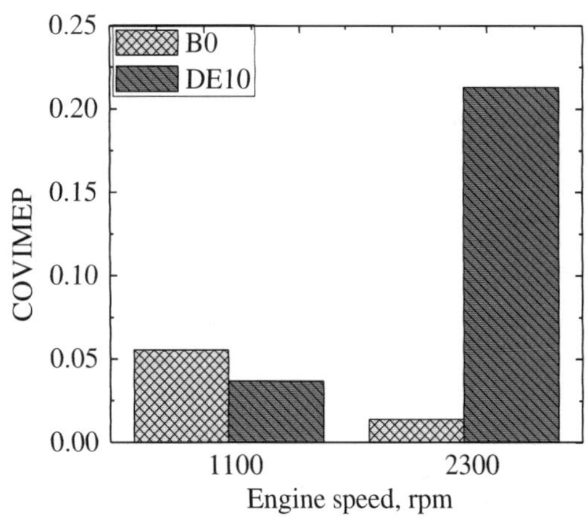

Figure 2. COV for IMEP (COVIMEP) versus full load at 1100 rpm and 2300 rpm

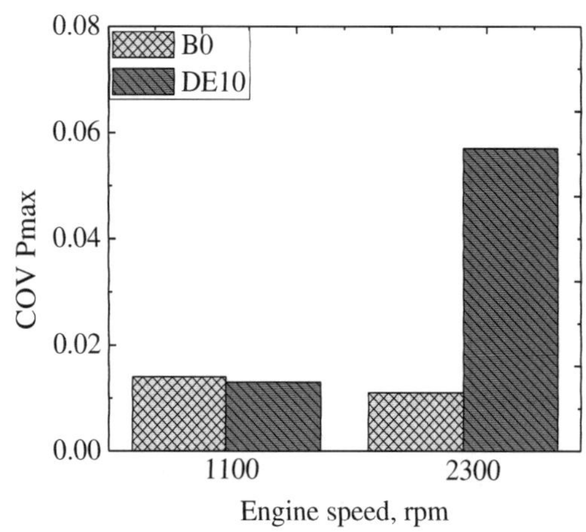

Figure 3. COV for maximum pressure, Pmax ($COVP_{max}$) versus full load at 1100 rpm and 2300 rpm

Comparing both two running conditions defined by 1100 rpm and 2300 rpm, at only DE10 fuelling, it was observed that at the speed of 2300 rpm, there was an increase of COVIMEP from 0.037% to

0.213% for DE10, as Figure 2 shows. However, the lower tendency of cyclic variations was observed for B0 respectively from 1100 rpm to 2300 rpm. The decreasing effect also appears for the B0 fuelling when the COVIMEP decreases from 0.056% to 0.014%. The cycle variability coefficient of maximum pressure, COVPmax, for DE10 fuelling increases from 0.013% to 0.057% at full load condition when the speed rises from 1100 to 2300 rpm, as Figure 3 presents. The cyclic variation is deteriorated at speed increasing, increasing with 338.5% for the DE10 fuelling. On the other hand, the $COVP_{max}$ for the B0 fuelling tends to decrease with 21.4% from 0.014% to 0.011% as the speed increases.

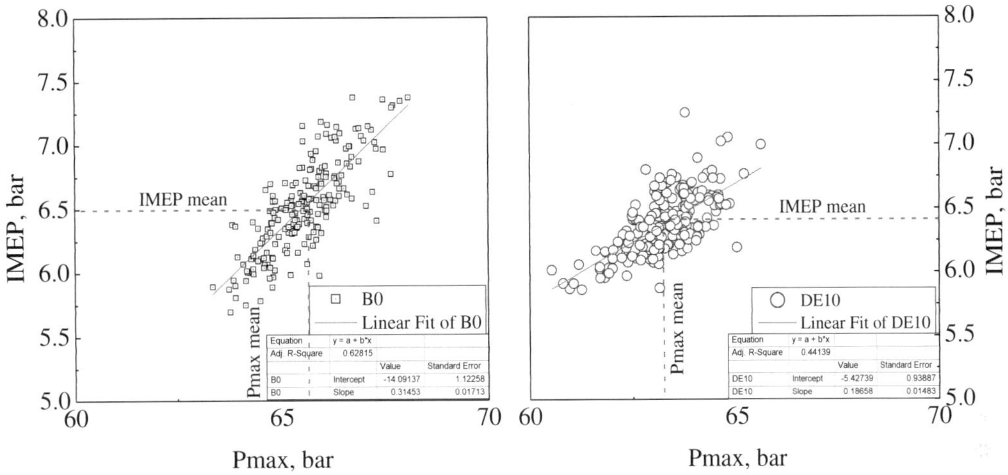

Figure 4. IMEP versus P_{max} with full load at 1100 rpm

Comparison for both B0 and DE10 in IMEP versus P_{max} in similar condition is illustrated in Figure 4. It was observed from the figure that maximum cylinder pressure traces for both B0 and DE10 vary for the 200 cycles. The higher intensity of cyclic variation was found for DE10, comparative to B0. The difference between the two pressure traces is further increases, not only due to the higher fuel mass at full load condition which is burned but also due to the faster combustion process at high engine load. Furthermore, combustion at high load with high engine speed starts earlier due to the more upper air and fuel masses, while remaining gas fraction that dilutes mixture is lower which benefits faster flame propagation.

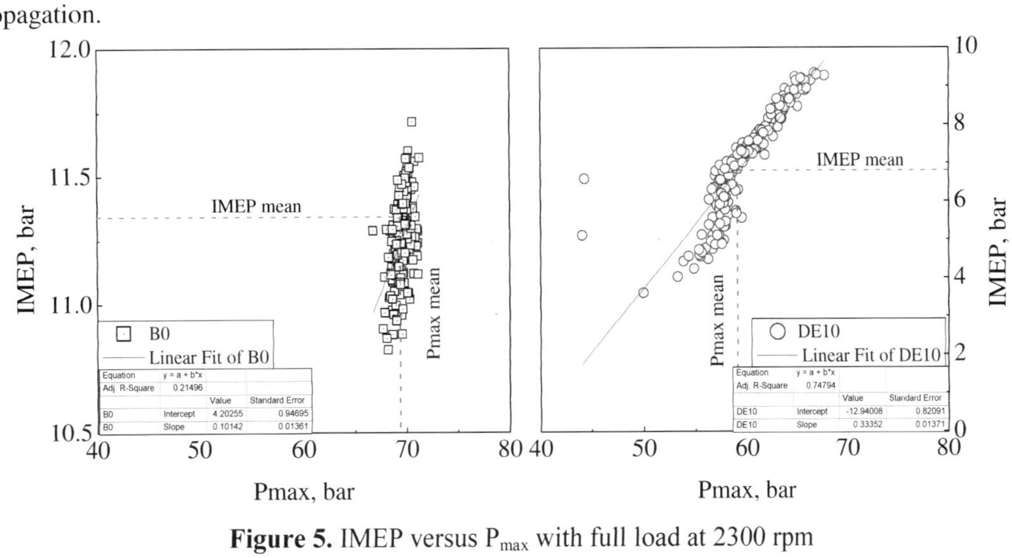

Figure 5. IMEP versus P_{max} with full load at 2300 rpm

The IMEP versus maximum pressure, P_{max} values for both B0 and DE10 increase when engine speed increases, as shown in Figure 5. It was observed from the figure that B0 has higher intensity for maximum pressure, P_{max}, comparative to DE10. Also, the maximum pressure, P_{max} for B0 increases as the engine speed increases from 1100 rpm to 2300 rpm. From the point of view, high maximum cylinder pressure cycles are associated with the fast burn rates owing to the increase in piston speed movement. Also, earlier initial flame kernel development leads to timely combustion evolution before TDC, that resulted in higher cylinder pressure. However, lower maximum cylinder pressure for DE10 was observed at 2300 rpm with the DE10 fuelling produced higher cyclic variation over 200 consecutive cycles.

4. Conclusion

The conclusions of the research work are listed as follow:

i. For DE10 fuelling at both speed conditions, the values of the coefficient of variation (COV) for indicated mean effective pressure (IMEP), COVIMEP increases comparatively to the values for only B0 fuelling, a fact which shows the improvement of the general response of the engine at full load condition.

ii. The value of $COVP_{max}$ for DE10 fuelling increases of 338.5% at full load condition when the speed increases from 1100 rpm to 2300 rpm. Also, for all speed conditions, the DE10 fuelling leads to the rise of cyclic variations for P_{max} value comparative to B0 fuelling. The increasing tendency shown for $COVP_{max}$ is related to the variation of COVIMEP and shows the improvement of the combustion process at DE10 fuelling.

iii. For the full load condition, the cyclic variations for B0 are improved at speed increasing comparative to the DE10 fuelling, which indicates that B0 has better combustion stability.

5. References

[1] H. Liu, B. Hu, and C. Jin, "Effects of different alcohols additives on solubility of hydrous ethanol / diesel fuel blends," *Fuel*, vol. 184, pp. 440–448, 2016.

[2] K. Nithyanandan, J. Zhang, Y. Li, H. Wu, T. H. Lee, Y. Lin, and C. F. F. Lee, "Improved SI engine efficiency using Acetone-Butanol-Ethanol (ABE)," *Fuel*, vol. 174, pp. 333–343, 2016.

[3] B. R. Kumar and S. Saravanan, "Use of higher alcohol biofuels in diesel engines : A review," *Renew. Sustain. Energy Rev.*, vol. 60, pp. 84–115, 2016.

[4] Y. Li, L. Meng, K. Nithyanandan, T. H. Lee, Y. Lin, and C. F. Lee, "Combustion , performance and emissions characteristics of a spark-ignition engine fueled with isopropanol- n -butanol-ethanol and gasoline blends," *Fuel*, vol. 184, pp. 864–872, 2016.

[5] B. M. Masum, H. H. Masjuki, M. A. Kalam, I. M. Rizwanul Fattah, S. M. Palash, and M. J. Abedin, "Effect of ethanol–gasoline blend on NOx emission in SI engine," *Renew. Sustain. Energy Rev.*, vol. 24, pp. 209–222, Aug. 2013.

[6] M. Mofijur, M. G. Rasul, J. Hyde, A. K. Azad, R. Mamat, and M. M. K. Bhuiya, "Role of biofuel and their binary (diesel-biodiesel) and ternary (ethanol-biodiesel-diesel) blends on internal combustion engines emission reduction," *Renew. Sustain. Energy Rev.*, vol. 53, pp. 265–278, 2016.

[7] A. Kujawska, J. Kujawski, M. Bryjak, and W. Kujawski, "ABE fermentation products recovery methods - A review," *Renew. Sustain. Energy Rev.*, vol. 48, no. August 2015, pp. 648–661, 2015.

[8] M. Lapuerta, J. Rodríguez-Fernandez, R. García-Contreras, and M. Bogarra, "Molecular interactions in blends of alcohols with diesel fuels: Effect on stability and distillation," *Fuel*, vol. 139, pp. 171–179, 2015.

[9] C. Wang, A. Janssen, A. Prakash, R. Cracknell, and H. Xu, "Splash blended ethanol in a spark ignition engine – Effect of RON, octane sensitivity and charge cooling," *Fuel*, vol. 196, pp. 21–31, 2017.

[10] A. Taghizadeh-Alisaraei and A. Rezaei-Asl, "The effect of added ethanol to diesel fuel on performance, vibration, combustion and knocking of a CI engine," *Fuel*, vol. 185, pp. 718–733,

2016.
[11] B. G. Moxey, A. Cairns, and H. Zhao, "A comparison of butanol and ethanol flame development in an optical spark ignition engine," *Fuel*, vol. 170, pp. 27–38, 2016.
[12] P. Schiffmann, V. Sick, and F. Foucher, "Multi-diagnostics Analysis of Flow Induced Combustion Variability at SI Engine-Like Conditions," *18th Int. Symp. Appl. Laser Imaging Tech. to Fluid Mech.*, pp. 776–796, 2016.
[13] S. Gurgen, B. Unver, and İ. Altin, "Experimental investigation on cyclic variability, engine performance and exhaust emissions in a diesel engine using alcohol-diesel fuel blends," *Therm. Sci.*, vol. 21, no. 1 Part B, pp. 581–589, 2017.

Acknowledgement

Universiti Malaysia Pahang is greatly acknowledged for the technical and financial supports under UMP Short grant (RDU172204). The authors appreciate the support of the management of Faculty of Mechanical Engineering, Universiti Malaysia Pahang for facilitating this research and UMP Central Laboratory for determining fuel properties.

IOP Conf. Series: Earth and Environmental Science **182** (2018) 012021 doi:10.1088/1755-1315/182/1/012021

A Study on an Absorption Refrigeration Cycle by Exergy Analysis Approach

Soheil Mohtaram[1], Wen Chen[*1], Ji Lin[1]

1 Institute of Soft Matter Mechanics, College of Mechanics and Materials, Hohai University, Nanjing 210098, China
Email: soheilmohtaram@gmail.com

Abstract. In this study, an absorption refrigeration cycle with the working fluid of water-lithium bromide is considered. The needful energy for generator is supplied by the steam at 100°C and in one atmospheric pressure. The exergy analysis is conducted on the whole cycle and it is calculated based on the first and the second laws of thermodynamics. Various components are compared in terms of thermodynamic efficiency. Finally the coefficient of the mentioned cycle is obtained. According to the simulation results, the highest rate of exergy destruction is in the absorber, and it is equal to 35.87 % of the destruction and the main cause of this irreversibility is heat transfer with the high-temperature difference. To improve this, we should increase heat exchange and then reduce the temperature difference. For the system performance improvement, particular attention should be paid to this part to reduce the outlet exergy.

1. Introduction

Limitations and prohibitions of the use of fluorocarbon gases have encouraged engineers to work more with absorption systems [1]. Exergy analysis [2]–[4] is of the methods widely used today to study engineering systems. This method combines the first and the second laws of thermodynamics [5] which relies mainly on the second law of thermodynamics, and serviceability. With the help of exergy analysis, it can be determined in which areas the level of serviceability is lost, based on the fact that the one can take actions to improve the system. Working fluid mixture for combined cycles can be divided into two categories: ammonia-water mixtures [6] and lithium-bromide [7]. Lithium bromide is a combination of an alkali metal salt (lithium) and a halogen in the form of white crystals which appears very similar to table salt (sodium chloride) dissolved in water and alcohol [8]. In the air, it will not be decomposed which owns a stable mix.

In this study, an absorption refrigeration cycle with the working fluid of water-lithium bromide with capacity of 500 tons is considered. The needful energy for generator is supplied by the steam at 100°C and in one atmospheric pressure. The exergy analysis is conducted on the whole cycle and it is calculated based on the first and the second laws of thermodynamics. Various components are compared in terms of thermodynamic efficiency, and then the coefficient of the mentioned cycle is obtained.

2. Description of Absorption Refrigeration Cycle and Assumptions

In Figure 1, a single effect absorption cycle is plotted. In an absorption system, condenser [9], evaporator [10]on the left breast and stifling cycle are the three components of the conventional vapor compression cycles. However, instead of compressors, four components are used in this paper: absorbers [11], pumps [12], expansion valves [13] and generators [14] as shown in Figure 1.

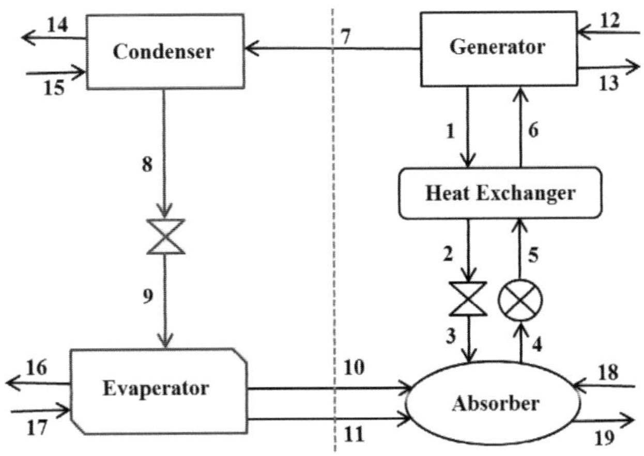

Figure 1. Schematic view of the absorption refrigeration cycle.

3. Exergy Analysis

Exergy analysis is a combination of the first and the second law of thermodynamics. We have the first law of thermodynamics for a sustainable flow-control volume as follows.

$$\dot{Q}-\dot{W}=\sum \dot{m}_{out}\left(h_{out}+\frac{V_{out}^2}{2}+gz_{out}\right)-\sum \dot{m}_{in}\left(h_{in}+\frac{V_{in}^2}{2}+gz_{in}\right) \quad (1)$$

$$S_{gen}=\sum \dot{m}_{out}\dot{s}_{out}\sum \dot{m}_{in}\dot{s}_{in}+\frac{\dot{Q}_{sur}}{T_0} \quad (2)$$

$$\dot{Q}_{sur}=-\dot{Q} \quad (3)$$

By eliminating Q in Eq. 1 and Eq. 2 and ignoring the entropy production (reversible), we can come to the exergy equation which is the maximum work from the beginning to the final state of the environment

$$\varphi=(\mathrm{h}-\mathrm{h}_0)-T_0(\mathrm{s}-\mathrm{s}_0)+\frac{V^2}{2}+gz \quad (4)$$

Exergy analysis assesses the system performance based on the exergy. Exergy is the maximum reversible work receivable from a system in transition from the initial state to a state of equilibrium with the environment.

For exergy of mass flow solution we have:

$$\varphi=\left[h_{(\mathrm{T,x})}-h_0\right]-T_0\left[s_{(\mathrm{T,x})}-s_0\right] \quad (7)$$

We calculated the exergy destruction in every component using the following equation:

$$\dot{E}_D=\dot{E}_{in}-\dot{E}_{out} \quad (8)$$

$$\dot{E}_D=\sum \dot{E}_{in}-\sum \dot{E}_{out}-\dot{Q}\left(1-\frac{T_0}{T}\right)-\dot{W} \quad (9)$$

We have examined the contribution of each component in exergy destruction with destruction percent relative to the total destruction cycle.

$$Y_{D,i}=\frac{\dot{E}_{D,\mathrm{i}}}{\dot{E}_{D,tot}} \quad (10)$$

Coefficient of performance (COP) expresses the energy ratio taken from cold water in the evaporator to the whole energy given to the system:

$$COP = \frac{\dot{Q}_e}{(\dot{Q}_e + \dot{W}_E)} \quad (11)$$

The second law efficiency expresses the ratio of useful exergy obtained from the system in evaporator to the exergy reported to the system in the generator.

$$E_{cooling} = \frac{\dot{E}_{15} - \dot{E}_{16}}{\dot{E}_{11} - \dot{E}_{12}} \quad (12)$$

Enthalpy of water and lithium bromide solution is calculated by Eq. 13. Moreover, solution entropy of data is taken from paper by Chow et al [8]. Enthalpy of lithium bromide and water solution according to the temperature and the concentration is calculated by the following equation:

$$h = E_1(x) + \mathrm{E}_2(\mathrm{x})\mathrm{T} + \mathrm{E}_3(x)T^2 \quad (13)$$

$$E_1(x) = -2024.18588321 + 163.2976010204x + 4.881268653177x^2 + 6.30250843\times10^{-2}x^3 -2.91350364\times10^{-4}x^4 \quad (14)$$

$$E_2(x) = 18.2816227619 - 1.169094163968x + 3.24785672\times10^{-2}x^2 - 4.03390218\times10^{-2}x^3 +1.85192774\times10^{-6}x^4 \quad (15)$$

$$E_3(x) = 3.70056321\times10^{-2} + 2.88756514\times10^{-3}x - 8.13075689\times10^{-5}x^2 + 9.91097142\times10^{-7}x^3 -4.44381071\times10^{-9}x^4 \quad (16)$$

In exergy calculations, unlike energy, there is no survival principle, and entry and exit of exergy do not match. We can calculate entry and exit of exergy to any system component by calculating exergy of the mass flow rate and through it; we obtain the destruction of exergy in each component. In Table 1, by comparing the exergy of different currents, we see that in places where water and lithium bromide flow, it has a considerable amount of more exergy that is due to solving two components (refrigerant and absorbent), and on the left side of the cycles only the refrigerant flows. On the right of the cycle, path (1) has the highest rates of exergy, which is the exit point solution with high-temperature generator.

According to the results in Table 2, the highest rate of destruction in absorber is 35.87 that have total destruction. The main cause of this is irreversibility is heat transfer with high temperature difference. To improve this, we can increase heat exchange and reduce temperature difference. This, on the other hand, increases initial cost. To improve system performance, particular attention should be paid to this part to reduce exergy exit. In addition, The Percentage of Exergy destruction of system components compared to total Exergy destruction of the system are shown in Figure 2.

Table 1. Characteristics of the Exergy cycle and calculated for each flow

Stream (i)	Temperature (C)	Pressure (Kpa)	Enthalpy (kj/kg)	Entropy (kj/Kg k)	Mass flow rate (kg/sec)	X (%LiBr)	Exergy (Kw)
1	98.67	8.687	249.0258	0.5058	9.046	64.6	929.87
2	58.3	8.687	176.257	0.3017	9.046	64.6	821.79
3	53.22	0.8756	176.257	0.3017	9.046	64.6	821.79
4	42.39	0.8756	117.775	0.2323	9.82	59.5	520.91
5	42.39	8.687	117.775	0.2323	9.82	59.5	520.91
6	76.83	8.687	183.892	0.432	9.82	59.5	585.78
7	93.3	8.687	2674.96	8.48	0.7756	-	118.21
8	43.11	8.687	180.45	0.6134	0.7756	-	1.67
9	5.056	0.8756	180.265	0.645	0.7756	-	-5.78

10	5.056	0.8756	2510.78	9.0255	0.7767	-	-131.91
11	5.056	0.8756	21.11	0.077	0.0189	-	0.05
12	100	101.325	2675.99	7.3556	1.112	-	543.23
13	100	101.325	419.012	1.3067	1.112	-	37.93
14	30	101.325	125.75	0.4365	77.1	-	13.03
15	36	101.325	150.822	0.5178	77.1	-	78.15
16	15	101.325	63.023	0.2236	52.53	-	46.55
17	7	101.325	29.508	0.1057	52.53	-	131.61
18	30	101.325	125.75	0.4365	112.97	-	19.09
19	35	101.325	146.44	0.504	112.97	-	84.05

Table 2. Destruction of Exergy and Exergy destruction percentage

Component	Input Exergy (Kw)	Output Exergy (Kw)	Exergy Destruction (Kw)	Exergy Destruction Ratio (%)
Absorber	709.03	604.96	104.07	35.87
Generator	1129.01	1086.01	43.00	14.82
Evaporator	40.77	-0.25	41.03	14.14
Heat Exchanger	1450.78	1407.58	43.03	14.89
Condenser	131.24	79.82	51.43	17.72
Refrigeration	1.67	-5.78	7.45	2.57
Solution Pump	520.935	520.91	0.025	0.000087
Overall System	3462.50	3172.34	290.17	100
		COP=0.7 , E=0.17		

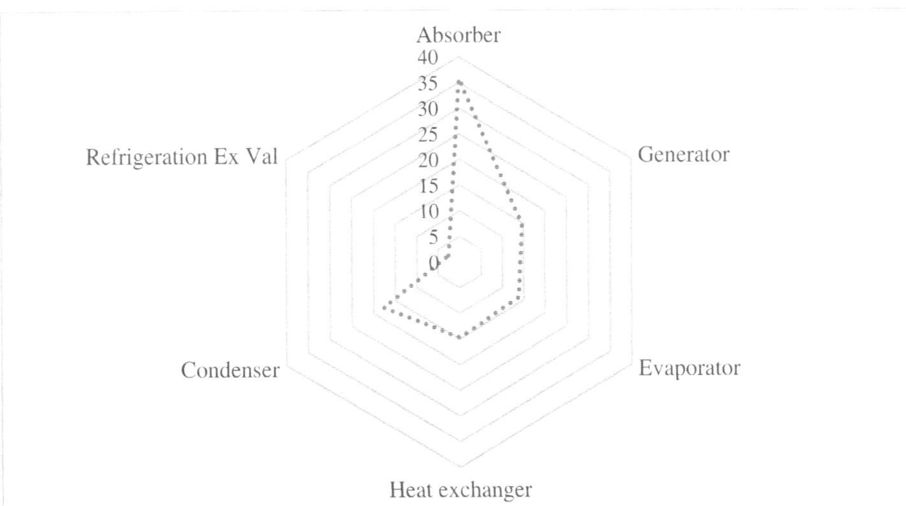

Figure 2. Percentage of Exergy destruction of system components compared to total Exergy destruction of the system

4. Conclusion

In this research, an absorption refrigeration cycle with lithium bromide as its working fluid is studied where the exergy analysis is conducted on the mentioned cycle. According to the simulation results, the highest rate of exergy destruction is in absorber and it is equal to 35.87 % of the total destruction. The main cause of this irreversibility is heat transfer with high temperature difference. To improve this, we should increase heat exchange and then reduce temperature difference. On the other hand, increases initial cost. To improve system performance, particular attention should be paid to this part to reduce exergy exit.

5. References

[1] M. Jelinek and I. Borde, "Single- and double-stage absorption cycles based on fluorocarbon refrigerants and organic absorbents," Appl. Therm. Eng., vol. 18, no. 9–10, pp. 765–771, 1998.
[2] S. Mohtaram, J. Lin, W. Chen, and M. A. Nikbakht, "Evaluating the effect of ammonia-water dilution pressure and its density on thermodynamic performance of combined cycles by the energy-exergy analysis approach," Mechanika, vol. 23, no. 2, 2017.
[3] S. Mohtaram, W. Chen, and J. Lin, "Investigation on the combined Rankine-absorption power and refrigeration cycles using the parametric analysis and genetic algorithm," Energy Convers. Manag., vol. 150, no. July, pp. 754–762, 2017.
[4] S. Mohtaram, W. Chen, T. Zargar, and J. Lin, "Energy-exergy analysis of compressor pressure ratio effects on thermodynamic performance of ammonia water combined cycle," Energy Convers. Manag., vol. 134, pp. 77–87, Feb. 2017.
[5] "Chapter 2 The first and second laws of thermodynamics," Int. Geophys., vol. 65, no. C, pp. 35–73, 1999.
[6] V. Zare, S. M. S. Mahmoudi, and M. Yari, "Ammonia-water cogeneration cycle for utilizing waste heat from the GT-MHR plant," Appl. Therm. Eng., vol. 48, pp. 176–185, 2012.
[7] O. Kaynakli, "Exergy analysis of absorber using water/lithium bromide solution," Heat Mass Transf. und Stoffuebertragung, vol. 44, no. 9, pp. 1089–1097, 2008.
[8] Y. Lu, Z. Tu, J. Shu, and L. A. Archer, "Stable lithium electrodeposition in salt-reinforced electrolytes," J. Power Sources, vol. 279, pp. 413–418, 2015.
[9] P. G. Youssef, S. M. Mahmoud, and R. K. AL-Dadah, "Numerical simulation of combined adsorption desalination and cooling cycles with integrated evaporator/condenser," Desalination, vol. 392, pp. 14–24, 2016.
[10] J. O. S. Parente, A. Traverso, and A. F. Massardo, "Saturator analysis for an evaporative gas turbine cycle," Appl. Therm. Eng., vol. 23, no. 10, pp. 1275–1293, 2003.
[11] N. I. Landy, S. Sajuyigbe, J. J. Mock, D. R. Smith, and W. J. Padilla, "Perfect metamaterial absorber," Phys. Rev. Lett., vol. 100, no. 20, 2008.
[12] and J. P. M. I. J. Karassik, W. C. Krutzsch, W. H. Fraser, Pump handbook, vol. 22, no. 6. 1976.
[13] N. J. Hewitt, J. T. McMullan, N. E. Murphy, and C. T. Ng, "Comparison of expansion valve performance," Int. J. Energy Res., vol. 19, no. 4, pp. 347–359, 1995.
[14] D. Bang and H. Polinder, "Review of generator systems for direct-drive wind turbines," Eur. Wind Energy, pp. 1–11, 2011.

Acknowledgment
This research investigation was supported by NSFC Funds (Nos. 11302069, 11372097), the 111 project under Grant B12032